Physiological Society Study Guides *Number 2*

Amino acid transport in animal cells

Physiological Society Study Guides

1 **Acid-base balance**
edited by R. Hainsworth *0 7190 1981 8 (hb) – 1982 6 (pb)*

2 **Amino acid transport in animal cells**
edited by D. L. Yudilevich and C. A. R. Boyd *0 7190 2432 3 (hb) – 2433 1 (pb)*

3 **The control of breathing in man**
edited by B. J. Whipp *0 7190 2463 3 (hb) – 2464 1 (pb)*

Physiological Society Study Guides *Number 2*

Amino acid transport in animal cells

Edited by
D. L. Yudilevich and **C. A. R. Boyd**

Manchester University Press

Published by Manchester University Press,
Oxford Road, Manchester M13 9PL, UK

British Library cataloguing in publication data
Amino-acid transport in animal cells. —
(Physiological Society study guides; no. 2)
1. Amino acids 2. Biological transport
3. Cell membranes
I. Yudilevich, D. L. II. Boyd, C. A. R.
III. Series
591.87′5 QP561

Library of Congress cataloging in publication data applied for

ISBN 0 7190 2432 3 *hardback*
ISBN 0 7190 2433 1 *paperback*

Typeset in Hong Kong
by Graphicraft Typesetters Ltd. Hong Kong

Printed and bound in Great Britain
by Billing & Sons Limited, Worcester.

Contents

List of contributors

C. A. R. Boyd Department of Human Anatomy, University of Oxford, South Parks Road, Oxford OX1 3QX, UK

H. N. Christensen Department of Biological Chemistry, The University of Michigan, Ann Arbor, Michigan, USA

A. A. Eddy Department of Biochemistry and Applied Molecular Biology, University of Manchester Institute of Science and Technology, PO Box 88, Manchester M60 1QD, UK

J. C. Ellory University Laboratory of Physiology, University of Oxford, Parks Road, Oxford OX1 3PT, UK

M. S. Kilberg Department of Biochemistry and Molecular Biology, University of Florida, Gainesville, Florida, USA

O. H. Petersen MRC Secretory Control Research Group, Department of Physiology, University of Liverpool, PO Box 147, Brownlow Hill, Liverpool L69 3BX, UK

D. L. Yudilevich Department of Physiology, King's College London, Campden Hill Road, London W8 7AH, UK

Preface

This book had its beginnings in a teaching symposium under the auspices of the Physiological Society, however it attempts to be more than the sum of a number of separate papers by various contributors on amino acid transport. In an Introduction the editors have attempted to acquaint the non-specialist reader both with terminology and with an overview of the subject. Basic further reading is suggested and the experimental approaches usually employed to investigate amino acid transport are emphasised. Throughout the book an updated version of what is known of the topic as well as an appraisal of future development is given.

There is a comprehensive list of references at the end of the book but each chapter is supported by a short list of further reading. The latter includes references which are of particular use for orientation within the topic.

Hence, to whom might this book appeal? To biochemists, cell biologists, physiologists, nutritionists, and also to those in more applied areas such as medical physiology and pharmacology. However, without doubt our main aim is to teach and hopefully to encourage students and younger scientists to enter the field; the Editors will then be more than pleased that the efforts of their contributors have not been in vain.

We are most grateful to the contributors and also to those colleagues (and their publishers) who gave us their kind permission to reproduce some of their work. The invaluable and patient secretarial assistance of Miss Jane Ballinger in Oxford and Mrs Ruth Okini in London is gratefully acknowledged.

Introduction

This book is about the passage of amino acids through membranes and begins with 'an historical introduction by H. N. Christensen who describes the development of experimental studies of amino acid transport from the standpoint of the work of his own laboratory over a period of 40 years. This chapter provides an overview of the topic, a framework for the following specialised chapters and an introduction to some general concepts. He begins with the early recognition of concentration gradients between cells and plasma, and goes on to describe studies of competition between amino acids and the identification of distinct transport systems for different groups of amino acids. He then relates elevated concentrations within cells to particular instances of rapid growth from which the idea of endocrine control of amino acid transfer necessarily follows. Finally the role of transmembrane gradients of Na^+ and K^+ ions in the 'energisation' of amino acid transport is introduced; these are 'energised systems which are regulated'. Hormonal control as well as cell responses to specific nutritional needs (excess or deprivation of amino acid supply) are important topics also covered in Chapter 1 by Christensen and M. S. Kilberg.

In Chapter 2, A. A. Eddy explores in a simple yet rigorous fashion the thermodynamics of co-transport of organic solutes with Na^+ ions – the basis of present understanding of the mechanisms of amino acid transport. These theoretical discussions of the energetics of model systems lead to the consequential importance of the membrane potential to amino acid transport and the effects of amino acid translocation upon membrane potential. These are examined in the particular context of pancreatic acinar cells by O. H. Petersen in Chapter 3.

Even though the function of amino acid transport systems is still not clearly understood in the most simple of cells, the erythrocyte, a number of carriers have been described. In Chapter 4 J. C.

Ellory reviews studies on the kinetics of amino acid transport in these cells which provide unique experimental material as they can easily be obtained from animals, including man, and hence provide important material for genetic studies of mutant amino acid transporters.

As tissues differentiate during development their phenotypic characteristics require differential gene expression and hence differential protein synthesis. Therefore amino acid requirements must vary from tissue to tissue and from time to time. Regulation of amino acid transport is required *in vivo* and this is modulated in a number of ways as discussed in Chapter 1. However, understanding how such interactions combine to match the requirements of a functioning organ is a complex topic discussed in Chapter 5. Here emphasis is placed on the polarity observed in epithelial cells of various organs, and particularly on the characteristics of transport at the basolateral side of epithelia or luminal side of endothelia as it relates to amino acid uptake from the blood into the cytoplasm and/or transcellular transfer of amino nitrogen.

Amino acids are found in living organisms, both in their free form and bound by amide linkage in peptides and proteins. There is great diversity of structure in the several hundred amino (and imino) acids now known to exist in nature (see Barret, 1985). However, this contrasts with the limited array of L-α-amino acids (nineteen) which are found in proteins and peptides of higher organisms. Figure 0.1 shows the structures of the 20 primary amino acids (including proline which is a cyclic α-imino acid). Natural amino acids are often referred to in abbreviated form by a convention of three letters. These abbreviations are given in Table 5.1, which illustrates characteristic concentrations in plasma and tissue of natural amino acids.

Although plants and some micro-organisms are capable of biosynthesis of all the primary amino acids, mammals lack the ability to synthesise a sufficient quantity of eight of these compounds, and these amino acids are called ‘essential’: isoleucine, leucine, lysine, methionine, phenylalanine, threonine, tryptophan and valine. Most species other than man are also unable to make histidine, and some arginine. In general, essential amino acids have to be obtained from the diet. Nutrition (and diet design) is obviously an applied field to which the concepts on

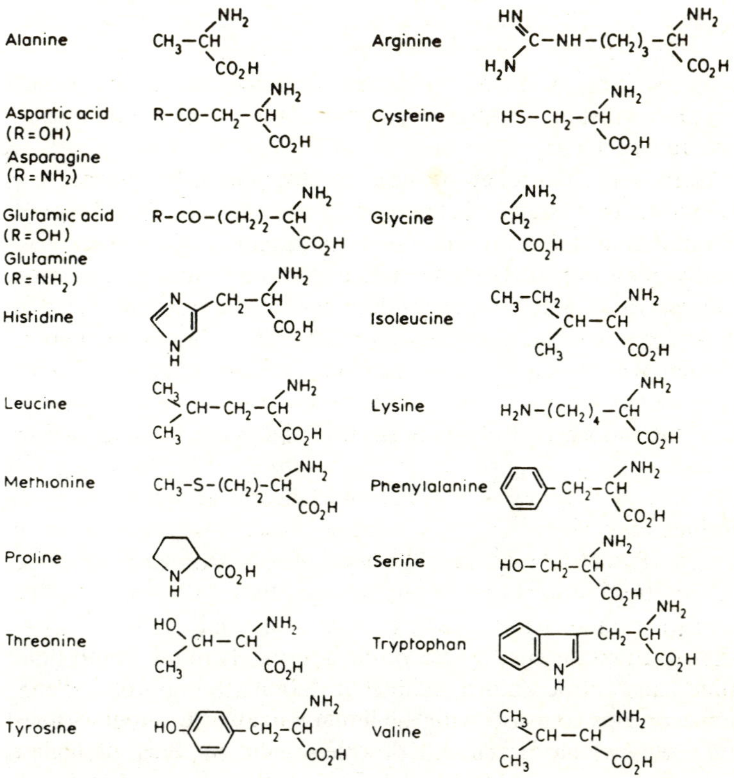

Fig. 0.1 The structure of the primary amino acids. The structures of these twenty amino acids, present in protein hydrolysates, were determined over a century: Braconett characterised glycine and leucine in 1820 and threonine was the last isolated in pure form, in 1925. (Taken with permission from 'The protein amino acids' by P. M. Hardy (1985). In: *Chemistry and biochemistry of the amino acids*. Barret, G. C., pp. 6–24. Chapman & Hall, London.)

amino transport given here will be of interest: for example, the effect of special diets on the regulation of amino acid uptake from the plasma into the brain (see Tews in Kilberg, 1986, and Bender in Barret, 1985).

In this book the reader will find extensive use of the word 'carrier', or 'transporter', or 'transport system', or simply 'system', to describe hypothetical proteins in the cell membrane. Such proteins will allow movement of an amino acid from one side of the cell membrane to the other.

Terminology in the field of biological transport may be at times appear confusing. This is not surprising since until the membrane proteins which are responsible for amino acid transport have been analysed at the level of atomic structure, the terminology employed to describe such transport systems will necessarily be based on indirect experimental evidence. Physiologists have traditionally divided transport mechanisms into two classes, 'carriers' and 'pores'. This division rests on criteria such as (i) the degree of molecular selectivity of substrate (high in the case of carriers which distinguish markedly, for example, between L and D isomers of amino acids); (ii) saturation of transport at high substrate concentration (typically found for carriers but not for pores); (iii) coupling of one flux to another (as for those carriers which show coupling of an amino acid flux to the flux of Na^+; see Robinson, 1975).

Our view of carriers has, however, altered from early concepts of shuttling 'ferry boats' to that of membrane proteins in which conformational changes in parts of the macromolecule allows the substrate binding site to be exposed alternately to the intracellular and to the extracellular medium. For pores, the notion that they act as narrow, water-filled channels permeable to those ions and molecules which fit through the hole has recently undergone a radical change largely because of the application of new techniques which have the ability to resolve single channel activity (see for example Chapter 3). This technical discovery has shown that the rate of passage of ions through one channel, when it is open, is greater than 10^6 ions per second. This is far too high for any mechanism other than a pore. Substrate selectivity, saturation and flux stoichiometry therefore no longer provide the best criteria for distinguishing between pores and carriers (see Hille, 1984). It should be emphasised that pores do not play a role in current ideas on the mechanism of amino acid transport.

The nomenclature 'cis' and 'trans' may also require explanation in reference to translocation of amino acids across membranes. These words are used operationally when a 'flux' is being measured to locate the compartment respectively from which (cis) and towards which (trans) the amino acid in question is moving.

Thus during influx experiments in single cells the cis compartment will be the extracellular and the trans compartment the intracellular fluid. During measurement of efflux of amino acid from such a cell, the terms naturally will have the opposite meaning with cis referring to the intracellular and trans to the extracellular medium. Transport across the plasma membrane must be distinguished from transport across a cell or a layer of cells, which involve movement of the substrate across membranes and cytosol.

To complicate matters, the student must also be aware that for transepithelial fluxes the terms cis and trans are also used. However in this case they refer to the respective extracellular fluids in contact with either face of the cell and here again sometimes confusing nomenclature is used. Since epithelial cells are known to have two poles from both morphological and molecular standpoints, the 'apical', or 'mucosal', or 'brush border' surface is compared to the 'basal', or 'serosal', or 'baso-lateral', or interstitial side.

A word about the relationship between the transport of amino acids into cells and their subsequent intracellular metabolic fate may also be helpful. Flows in biological systems may be considered either in terms of atoms (for example of carbon as it passes down a given metabolic chain through a series of metabolic intermediates) or in terms of flows of the same molecule across barriers separating aqueous compartments (for example across the plasma membrane or across the membrane of various intracellular organelles). In both instances proteins act as the catalysts (whether as 'enzymes' or 'transporters') and in the living cell both types of catalysis naturally act together in a concerted fashion. Importantly, therefore, either transport or metabolism may be the slowest (rate limiting) step to the cell handling of a particular amino acid or indeed any other metabolic substrate. Furthermore, both enzymatic and transport activity may be characterized by the kinetic constants (maximal velocity, V_{max} and affinity constant K_m) using the Michaelis-Menten equation. (The reader should consult Stein, 1986, for some useful and sophisticated developments in the field of modelling and mathematical formulation of the complex kinetics required for detailed study of transport phenomena.)

The student should not neglect the multiple metabolic reactions involving amino acids when considering biological aspects of membrane transport (see, for example, Newsholme & Leech, 1983, and Barret, 1985). An example of pharmacological importance occurs in the study of transport of tryptophan (incidentally, this is the only amino acid which in plasma is not in the free state since it avidly binds to plasma albumin). This aromatic amino acid is an essential amino acid which, apart from its role in protein synthesis, is a precursor for a number of biologically active compounds, including the neurotransmitter amine 5-hydroxytryptamine (serotonin). Also phenylalanine and tyrosine are amino acids directly involved in the synthesis of noradrenaline. An attempt to identify whether transport or metabolism is the rate limiting step for utilisation (gluconeogenesis) for each of a number of amino acids, in liver, is presented in Chapter 5, section 5.10.

This is not a book that will tell the reader all about how to set about investigating amino acid transport experimentally. However, since our aim is to stimulate experimental research, we have attempted to summarise in Table 0.1 the experimental strategies which may be adopted in the laboratory. The review by Lerner (1985) will orientate the reader to the literature and background of this apparently complex Table. As an example of one of the experimental approaches to amino acid transport consider the membrane potential. The potential as discussed in Chapters 2 and 3 contributes to the driving force for Na^+ movement across the cell membrane and this may be altered experimentally by a number of strategies which are indicated in Table 0.1. Either the concentrations of permeant ions outside the cell may be changed or the permeability of the membrane may be altered by the use of chemicals (ionophores) which specifically increase the permeability of the cell to one ion (for example the ionophore valinomycin increases K^+ permeability). By such means the membrane potential can be manipulated in a precisely defined fashion. Another experimental approach may involve the manipulation of the intracellular energy supply. ATP production may be inhibited by poisons of glycolysis such as cyanide or of oxidative phosphorylation such as dinitrophenol (DNP). Alternatively oxidative metabolism may be suppressed in a nitrogen atmosphere. Transport may then be studied under conditions of ATP depletion. When employing protein synthesis inhibitors these may influence transport either by the inhibition of synthesis of new

Table 0.1 *Experimental manipulations affecting amino acid transport*

Electrochemical gradient of ions:
Chemical gradient
Extracellular sodium and other ions (K^+, Li^+, H^+, Cl^-)
Intracellular sodium: sodium pump inhibition (ouabain, zero-external potassium)
Electrical gradient
Passage of electrical current
Gradients of permeable ions (K^+, Cl^-)
Ion permeability changes. *Ionophores*: valinomycin, CCCP, nigericin, gramicidin, monensin, neomycin. *Channel blockers*: amiloride, TTX, quinine. *Channel activators*: intracellular calcium, propranolol

Metabolism:
Intracellular metabolites
Glycogen, glucose, metabolic intermediates
ATP depletion
Anoxia, low temperature, DNP, cyanide, arsenate

Humoral agents:
Hormones
Insulin, glucagon, growth hormones, vasopressin, gastrin, cholecystokinin, steroid hormones (dexamethasone), thyroid hormones
Neurotransmitters
Acetylcholine and cholinergic agonists (carbachol)
Catecholamines: alpha- and beta-adrenergic agonists (noradrenalin and isoproterenol)
Paracrine
Prostaglandins and synthesis inhibitor indomethacin

Intracellular second messengers:
Cyclic nucleotides cAMP and DbcAMP, cGMP and 8-BrcGMP
Intracellular calcium Ca^{2+}-ionophore (A23187)

RNA synthesis inhibitors: actinomycin

Protein synthesis inhibitors: cycloheximide, puromycin

Cytoskeleton disruption: *Microfilaments* (cytochalasin B)
Microtubules (colchicine)

Sulphydryl-group inhibitors: NEM, mercuric compounds (PCMBS)

Heterogenous inhibitors: phloretin, aspirin, bile salts, nicotine, ethanol

TTX = tetrodotoxin;
CCCP = carbonyl cyanide *m*-chlorophenylhydrozone;
DNP = dinitrophenol;
NEM = N-ethylmaleimide;
PCMBS = *P*-chloromercuriphenylsulphonate;
DbcAMP = dibutyryl cyclic AMP;
8-BrcGMP = 8-bromo cyclic GMP.

carriers or by interrupting the relationship between amino acid transport and protein synthesis. Induction of one of the specific transport proteins, the so-called A system, has been achieved by either hormones (insulin, glucagon) or by substrate starvation (i.e. lack of certain specific amino acids). Experimentally the induction of system A in cells cultured in amino acid-free medium has been inhibited by protein synthesis and/or RNA synthesis inhibitors (Table 0.1).

There is only limited reference in this book to the genetics of amino acid transporters. However, it should be realised that 'experiments of nature' are increasingly contributing and sometimes overtaking experimental investigations. This must provide another exciting area for future advances. Table 5.2 illustrates some of the genetic defects which are known to affect amino acid transport in the kidney of human and other animals. Although distinct genetic defects affecting either amino acid transport proteins or enzymes of amino acid metabolism have been described, indirect interactions between transport and metabolism emerge (e.g. as found in phenylketonuria). Orientation will be found in the review by Willner & Meissner (1981).

In 1987 the topic of this volume, amino acid transport, finds itself, as do many other important areas of cell physiology, at a crossroads. The application of molecular biology, although in its infancy, promises to contribute highly to this field in the next decade, and the structural knowledge of the membrane proteins which will be discovered by such approaches should aid in the understanding of the molecular mechanisms involved in amino acid translocation. At present, studies in bacterial systems lead the way (see Wright *et al.*, 1986) but substantial advances in animal cells (though not for amino acid carriers) have been described in the last 18 months (e.g. Mueckler *et al.*, 1985, on the glucose transporter; Kopito & Lodish, 1985, on the anion carrier; and see Racker, 1985).

Further reading

(Other literature cited is listed in the compiled Reference list at the end of the book.)

Barret, G. C. (ed.) (1985). *Chemistry and biochemistry of the amino acids*. Chapman & Hall, London.

Christensen, H. N. (1975). *Biological transport*, 2nd Edn. W. A. Benjamin, Reading, Mass.

Eckert, R. & Randall, D. (1983). *Animal physiology*, 2nd Edn. W. H. Freeman, New York.

Harrold, F. M. (1986). *The vital force; a study of bioenergetics*. W. H. Freeman, New York.

Harrison, R. & Lunt, G. G. (1980). *Biological membranes, their structure and function*. 2nd Edn. Blackie, Glasgow.

Hille, B. (1984). *Ionic channels of excitable membranes*. Sinauer, Sunderland, Mass.

Hofer, M. (1977). *Transport across biological membranes*. Pitman, London.

Lerner, J. (1985). Effectors of amino acid transport processes in animal cell membranes. *Comparative Biochemistry and Physiology*, **81A**, 713–39.

Newsholme, E. A. & Leech, A. R. (1983). *Biochemistry for the medical sciences*. John Wiley, Chichester.

Nicholls, D. G. (1982). *Bioenergetics, an introduction to chemiosmotic theory*. Academic Press, London.

Racker, E. (1985). *Reconstitution of transporters, receptors, and pathological states*. Academic Press. New York.

Robinson, J. R. (1975). *A prelude to physiology*. Blackwells, Oxford.

Stein, W. D. (1986). *Transport and diffusion across cell membranes*. Academic Press, London.

Willner, D. & Meister, A. (1981). A survey of inborn errors of amino acid metabolism and transport in man. *Annual Review of Biochemistry*, **50**, 911–68.

1

Amino acid transport across the plasma membrane: role of regulation in interorgan flows

H. N. Christensen Department of Biological Chemistry
The University of Michigan Ann Arbor *and* **M. S. Kilberg**
Department of Biochemistry and Molecular Biology
University of Florida.

1.1 **An introduction to amino acid distribution**

1.1.1 *Gradients between cells and the extracellular fluid*

Van Slyke & Meyer (1913–1914) used their new nitrous acid method to show that the free amino acids were at much higher concentrations in tissues than in plasma of the dog, and that the injection of an amino acid mixture (hydrolysed casein) raised the tissue levels far more than it did the plasma levels. A quotation from these authors deserves mention.

> The amino acids of the blood appear therefore to be in equilibrium with those of the tissues, a condition which accounts for all the observed phenomena, and would also account for any transfer of amino acids which may occur from organ to organ, or from maternal organs to be fetus.

The only change one could urge in the light of decades of progress might be to replace the word *equilibrium* with a term more suitable for describing an energised transport system which is regulated.

1.1.2 *Transport competition by other amino acids: a classification of transport carrier systems*

In early experiments, we perturbed the distribution of glycine and glutamine by feeding single amino acids to guinea pigs by stomach tube (Christensen *et al.*, 1948a). An hour later we analysed plasma, liver and muscle for these two amino acids, and the total amino acids yielding CO_2 with ninhydrin. Our results showed what today we would expect from amino acids competing for transport among the specific transport systems shown in Table 1.1. (There is in Chapter 4 and Figure 4.1 further useful illustration of the classification of amino acid transport systems.) The high distribution ratio (cellular to extracellular) was lowered by feeding any of several L amino acids (shown for glycine distribution in Figure 1.1*a*). Reciprocally, the large increases in plasma glycine produced by feeding that amino acid lowered the tissue: plasma ratio for the other amino acids to 44% of the control. In our 1948 paper we wrote:

> The amino acids appear to show competitive inhibition among themselves for the means by which the cells concentrate amino acids, indicating that these means do not operate independently for each amino acid. . . . it may be inferred that when a strongly asymmetric accumulation of amino acids is produced by feeding or injection, the ability of cells to retain other amino acids will be handicapped, and one may expect an inhibitory effect on growth or N retention.

1.1.3 *Amino acid sequestering produced by transport competition*

What we failed to emphasise properly in 1948 can also be seen in Figure 1.1*a*: phenylalanine, leucine and tryptophan tended moderately to *elevate* the distribution of glycine between tissue and plasma. This paradoxical result could no longer be ignored when examining more closely the perturbations of the glycine gradients for muscle (also reported in our 1948 paper) by means of a graph such as in Figure 1.1*b* (Christensen, 1986). Only when realising that we had been dealing with two distinct populations of amino acids were we forced to face this paradox. One population, including alanine, serine, threonine and proline, showed the expected competitive inhibition of uphill transport of glycine (Figure 1.1*b*). Another population, including tryptophan, leucine, phenylalanine, tyrosine, methionine and histidine had to our surprise *intensified* the gradients for our test amino acid, glycine.

The reason for this behaviour becomes clear in the retrospective light of the discovery of the combination of distinct transport systems that we now recognise as serving for cellular uptake of amino acids (Table 1.1; see Christensen, 1985). In the ten experiments of Figure 1.2 that show lowered glycine gradients, the perturbations were produced by amino acids that are typically

Table 1.1 *Abbreviated summary of some known transport systems in tissues and cells of higher animal* (references in Christensen, 1982)

Systems for dipolar amino acids

Na^+ dependent systems:

System Gly	Glycine and sarcosine. Widespread	Characteristic tolerance of an N-methyl group (System Gly and System A)
System A	Most dipolar amino acids. Ubiquitous	
System ASC	Numerous dipolar amino acids but not their N-methyl derivatives. At lowered pH, certain anionic amino acids	
System $B^{0,+}$	Especially wide in scope, as defined in Section 1.2.3	
System N	Histidine, glutamine, asparagine. Hepatocytes, distribution otherwise not determined.	
β-System	β-alanine, taurine, β-aminobutyric acid	

Na^+ independent systems:

System L[a]	Prefers dipolar amino acids with bulky side chains Possible variants must be considered
System T[a]	Prefers tryptophan to leucine, etc. Human red cells
System asc	Prefers threonine, serine, alanine, cysteine Erythroid cells, but usually not in mature mammalian red cells. Analogy to system ASC has not proved common origin

Systems for anionic amino acids

X_A^-, X_G^-, X_{AG}^-, X_C^-	Identify systems preferring respectively aspartate, glutamate, neither, or cystine and glutamate Relation to neurotransmittory actions?

Systems for cationic amino acids

y^+	Lysine, arginine, ornithine Two forms may appear together in gut. System specific to ornithine known in hepatic mitochondria

[a]With the two historical exceptions L and T, lowercase letters are preferred for designating Na^+ independent transport systems (Bannai *et al.*, 1984).

(Editor's note: See Figure 4.1 and Chapter 4 for further illustration of the various transport systems, and an additional system, C, there proposed. Also useful are Table 1 in Christensen (1982), Table 1 in Kilberg (1982), and Table 1 in Bannai & Tateichi (1986).)

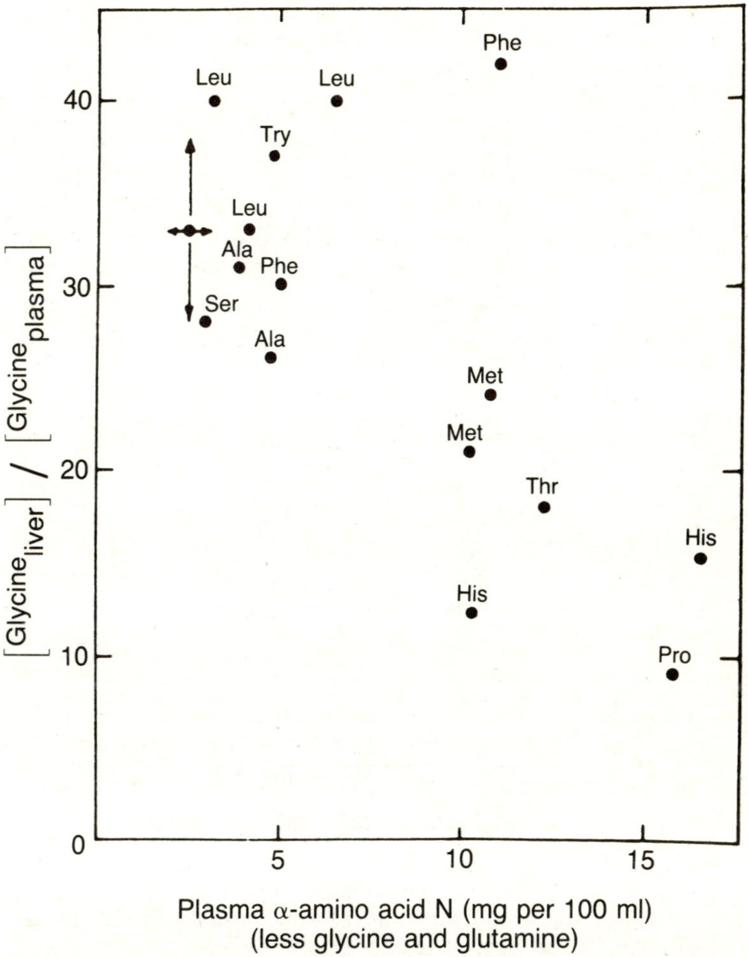

Fig. 1.1(*a*) The effect of administering a single amino acid on the distribution of glycine between tissue and plasma. Each point represents the results obtained when a fixed dose of a given individual amino acid, as indicated, was given to a guinea pig by mouth. The glycine concentration in liver and plasma was determined and plotted as a function of the resulting plasma α-amino acid N concentration. Tissue was taken for analysis at 3 h. (Taken with permission from Christensen, H. N. *et al.* (1948). *Journal of Biological Chemistry*, **172**, 515–24.)

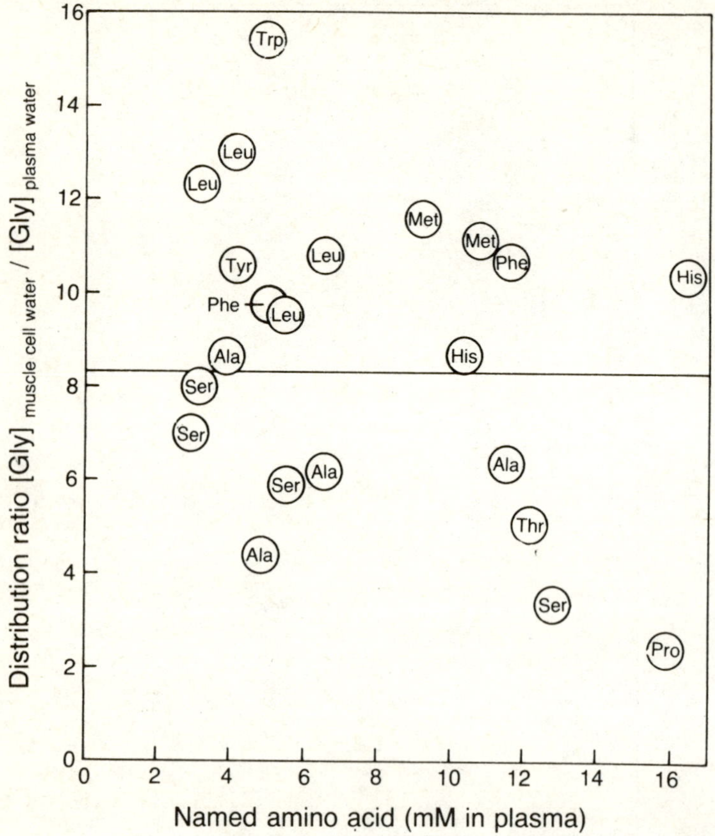

Fig. 1.1 (*b*) The influence of various individual amino acids on the distribution of glycine between thigh muscle and plasma in the guinea pig. The effects of the amino acids fall into two populations: those that decrease the glycine distribution ratio (below the line), and those (high reactivity with system L), which intensify the muscle accumulation of glycine (above the line). Note that these latter amino acids all have high reactivity with system L. (Redrawn from Christensen, N. N. *et al.* (1948). *Journal of Biological Chemistry*, **172**, 515–24.)

transported by the highly concentrative, Na^+ dependent system A (Oxender & Christensen, 1963). In contrast the six amino acids which intensified the glycine gradient are transported to major degrees by system L. The important concept represented here is shown in Figure 1.2.

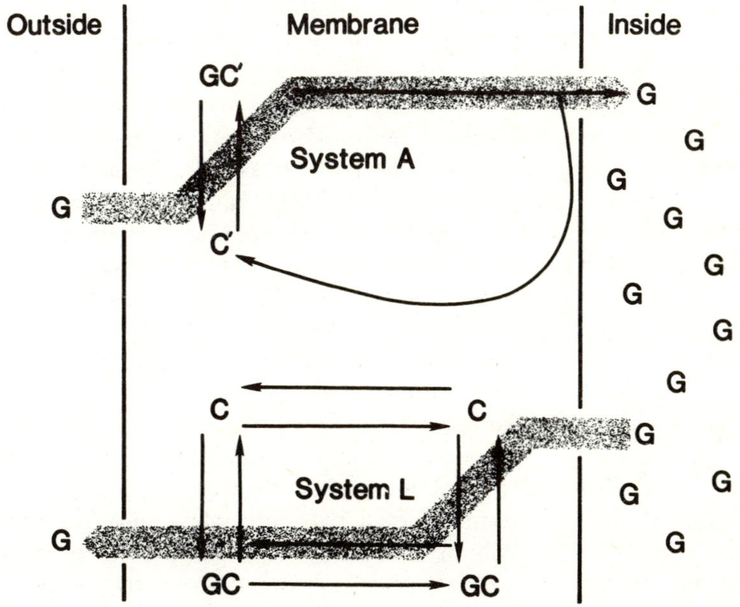

Fig. 1.2 Diagram to illustrate that an amino acid may be pumped into a cell by one strongly energised transport system (here system A) and released from the cell by another (here system L). Attention is drawn to the action of phenylalanine and analogous amino acids in inhibiting exodus from liver, muscle, or other tissues via system L, of amino acids entering largely by system A. Note that each of the two transport systems illustrated is able to produce net transport in either direction, although system A is much less reversible. The operation of system L leads in addition to conspicuous exchange of its substrates across the plasma membrane. (Modified after Christensen, H. N. (1975). *Biological transport*, 2nd edn. W. A. Benjamin, Reading, Massachusetts.)

System A has, as illustrated, a very low reversibility; it tends to generate gradients of many neutral amino acids which are too steep to be sustained by Na^+ independent System L. Even though the latter system can be shown to operate uphill in the inward direction under selected experimental conditions, it is regularly forced to operate in reverse by the steep gradients generated by the more steeply uphill Na^+ dependent transport systems. Accordingly an amino acid supplied in excess can exert its transport-perturbing effect either on the net tissue uptake or on the net release of an endogenous amino acid, depending on the net directional role of the transport system suffering the greater competitive interaction between the pair of amino acids selected for test. Christensen & Cullen also showed that two model amino acid substrates for system A lowered the hepatic levels of various amino acids relative to their plasma levels (see Christensen, 1982; 1984). By selecting instead a bicyclic amino acid analogue designed to inhibit System L, we showed intensified hepatic gradients of a number of amino acids, again corresponding rather well to the transport roles to be expected for the several transport systems. Interestingly the failure of phenylalanine and tyrosine to lower the glycine distribution ratio (Figure 1.1*b*) suggests that these amino acids may mainly be transported into liver by the recently described System T (refer to Chapter 4).

Recently it was proposed further (Christensen, 1986) that the paradoxical lowering of a number of the plasma amino acids repeatedly reported in patients with phenylketonuria can be explained by the inhibitory action of the large accumulations of phenylalanine, not its now recognised action on the entry of certain amino acids into the brain which Christensen suggested in 1953, but rather on the net efflux of a somewhat different group of amino acids from liver, muscle and other tissues. The inhibition of their release from such tissues by excess phenylalanine (and probably also by excess leucine in maple syrup urine disease) appears to lead to their sequestration in these tissues, hence presumably also to acceleration of their catabolism, and thus to a second mode by which transport competition impoverishes the amino acid nutrition of the developing brain, namely by reducing the mass action of various amino acids serving to drive their transport across the blood–brain barrier. Because System L operates mainly inward across the blood–brain barrier (except probably for glutamine; James *et al.*, 1979) and for a number of amino acids outward from other tissues, its inhibition by phenylalanine or leucine excess appears to operate in both cases unfavourably to the amino acid nutrition of the brain (see Chapter 5).

1.1.4 *Growth and amino acid transport*

Our laboratory followed (1948) with examples of the association of rapid growth and elevated cell concentration of amino acids.

In **fetal tissues** the levels of amino acids were shown to be elevated over those of the maternal muscle by two factors: the interpolated concentrative transport across the placenta, and an intensified uptake, at least by the fetal muscle *per se* compared with the maternal muscle (Christensen & Streicher, 1948 – see Christensen, 1984). In this paper we attributed the lowering of the plasma amino acids during human pregnancy to 'the avid removal of amino acids by the placenta'. We also observed that mock surgery (serving as a control procedure for the hepatectomies described in the following paragraph) lowered the plasma concentration of amino acids, an effect we attributed to stimulation of hepatic amino acid uptake associated with the previously demonstrated acceleration of protein catabolism. Thus for selected instances we attributed declines of the plasma amino acid levels to intensified amino acid withdrawal by tissues, and conversely attributed elevated plasma amino acids to handicapped tissue accumulation. These attributions, perhaps because they reverse the usual didactic association of elevated circulating amino acid levels with enhanced nutrition, have had slow acceptance, although we have not found cause to doubt their role in the interpretation of plasma amino acid levels. The concentrative transfer of glycine across the guinea pig placenta could, we found, be sharply handicapped by feeding extra histidine, methionine or proline (see placental transport in Chapter 5).

After subtotal hepatectomy in the rat, the residual portion of the liver begins to accumulate amino acids more strongly in approximate coordination with the beginning of **mitosis** and **hepatic regrowth**.

The unusual concentrative powers of **neoplastic tissue**, demonstrated first for the Ehrlich ascites tumour cell (cf. Chapter 2), permit a tumour to capture amino acids from normal tissues of the host.

The influence of **tissue ageing** was emphasised in our 1948 paper. Twitty & van Wagtendonk (1940) reported that an eye transplanted from a younger to an older larval *Ambystoma tigrinum* subsequently grew faster than its untransplanted mate remaining upon the younger salamander. 'Had transplanting the eye placed it in a richer medium?' The authors proceeded to cite their analyses showing that such was emphatically the case. '. . . the blood amino acid concentration increased with the age of the larvae as a result of the decline of the growth potentialities of the organs.' In agreement with this idea, Christensen's laboratory subsequently found (Riggs *et al.*, 1958) decreasing 'amino-acid hunger with age' allowing elevation of plasma levels.

1.1.5 *Endocrine control of amino acid transport*

We subsequently focused more specifically on the endocrine control of amino acid transfer, and introduced the use of metabolism-resistant model amino acids (in this case 1–[^{14}C]–2–aminoisobutyric acid, AIB) for investigating hormonal control (Noall *et al.*, 1957). Well over a thousand papers have appeared subsequently in which AIB has been used in this way. In this 1957 paper we viewed hormonal control as a regulation applied by superimposition on a primitive membrane transport activity. We examined endocrine changes as modifying the fate of the amino acids in the higher animal by changing the proportion of the amino acids serving anabolically and for growth, relative to the proportions used catabolically. In this study, we added evidence for hormonal control of membrane transport following hormone administration *in vivo* in the rat, thus:

(i) adrenaline administration intensified the accumulation of ^{14}C-AIB by various tissues;
(ii) pituitary growth hormone intensified the accumulation of [^{14}C]-AIB by all tissues studied except the heart;
(iii) hydrocortisone intensified the accumulation of [^{14}C]-AIB by liver, but not by skeletal muscle, heart, or duodenum;
(iv) oestradiol led, 20 h later, to an approximate tripling of the [^{14}C]-AIB in the uterus but not in the liver; and
(v) insulin *in vitro* stimulated AIB capture by the isolated, intact rat diaphragm (Kipnis & Noall, 1958).

The five hormonal effects on amino acid transport thus demonstrated have subsequently been studied at various degrees of increased sophistication, and other endocrine effects on membrane transport added to the list, in an updating of that subject in section 1.3 of this chapter.

1.1.6 *Energisation of amino acid transport*

Christensen's first decade of the study of amino acid transport into tissues and cells of the higher animal included studies which supported a crucial role of the alkali metal ions in the concentrative uptake of glycine, and which showed reciprocal movements of Na^+ and K^+. The energization could be due to gradients of either Na^+ or K^+, or of both. Despite justified reservations and an occasional, incompletely-justified bias for K^+, we insisted (Riggs *et al.*, 1958) that:

> A propelling function might instead be assigned to the Na^+ movements, which are reciprocal to the K^+ movements. In the same way consideration might be narrowed to a bound form of Na^+ *influx*; that is, the excess of Na^+ influx could occur in the form of a complex between the carrier, glycine, and sodium ion.

This version of the hypothesis has been widely accepted, although ironically K^+ movement continue to play a role, albeit indirectly, in amino acid transport. (The energisation of amino acid transport is discussed in Chapter 2.) Interconversions of transmembrane gradients have subsequently assumed a key role in membrane bioenergetics, as marked by the 1978 Nobel award in chemistry to Peter Mitchell (see Mitchell, 1977).

1.2 **Regulation of amino acid transport at the cellular level***

1.2.1 *Reticulocyte maturation*

The rabbit reticulocyte contains glycine at levels three to five times higher than the plasma. Furthermore, these cells take up glycine against concentration gradients, whereas no obvious gradient was maintained or established by the mature red blood cell (Riggs *et al.*, 1952). The uptake of glycine by reticulocytes was correspondingly more sensitive to respiratory poisons. More recently our laboratory showed that transport systems regress at different rates during reticulocyte maturation. System L persists into maturity largely undiminished, as it does in the human red cell. In contrast to these studies on rabbit erythrocytes uptake by system ASC

---

* Studies on the regulation of amino acid transport at the cellular level have been concerned with identifying which of the particular systems (Table 1.1) are present at different stages of the development of particular cells, which persist, and the rates at which they regress with maturation or differentiation. Another type of regulation is the feedback mechanisms which occur in conditions of excess amino acids or the opposite condition of amino acid starvation, i.e. adaptive regulation (see pp. 23–35). (Editor's note.)

although declining nevertheless remains appreciable through much of the life of the human red blood cell. We looked for spontaneous regression *in vitro*, but saw only a small increase in System ASC transport during 6 h of incubation of rabbit reticulocytes. Stimulation by the concomitant amino acid starvation could perhaps have masked any effect of ongoing reticulocyte differentiation. These results emphasise the importance of the discrimination of distinct transport systems (Table 1.1) in the study of their development and regulation (Christensen, 1975; 1985).

Study of transport systems surviving maturation of the sheep (Young *et al.*, 1975) and horse (Fincham *et al.*, 1985) erythrocyte, particularly the discovery of contrasting phenotypes of the same species in which survival is defective, allowed the discovery of a Na^+ independent transport system with some striking similarities to Na^+ dependent system ASC in its amino acid selectivity. Because of these similarities, and the variable persistence of System ASC during erythrocyte maturation, it was proposed that the new system (System asc) might arise by the loss of the Na^+ dependency of System ASC. Through the discovery of a quite similar system in the pigeon erythrocyte, a nucleated cell analogous to precursors of mature mammalian erythrocytes and a cell that retains high amino acid transport, the alternative hypothesis emphasises that the new Na^+ independent system of Young and colleagues is instead a persistent biological entity, and not merely a product of the terminal differentiation of mammalian red cells (Vadgama & Christensen, 1985). Subsequently we have found a corresponding Na^+ independent transport system also conspicuous in erythroid cells separated by density centrifugation from the fetal rat liver. (Recently system asc has been identified in the rat exocrine pancreatic epithelium; Mann & Peran, 1986.) In agreement with one of the conventions accepted by a number of investigators concerned with discrimination of amino acid transport systems, workers in this field have agreed to designate the new transport component as system asc, the lower case letters recognising it conventionally (see Bannai *et al.*, 1984) as a Na^+ independent analogue to system ASC. Discovery of further genetic variations in the persistence of transport systems during differentiation of mammalian erythrocytes may be expected, because failure of their persistence might have only minor influences on the survival and function of the circulating erythrocyte, provided only that any special metabolic needs retained by the mature cell are met, such as the capture of cysteine and other precursors of glutathione. (See Chapter 4 for further discussion on transport of amino acids in the erythrocyte.)

1.2.2 *Fetal and mature hepatocytes; hepatoma cells*

Large changes in the amino acid transport systems occur in the rat hepatocyte, as observed under primary culture, between its perinatal state in the fetus and its mature state in the adult rat. These changes include the regression of system y^+ for cationic amino acids. It is interesting that this system reappears, however, on transformation as seen in certain hepatoma cell lines (White & Christensen, 1982) and can be induced by treatment with glucagon or by experimental pancreatic diabetes (Handlogten & Kilberg, 1984). Furthermore, system A appears to be replaced at adulthood by a variant system similar to that seen in fetal life except in details of substrate specificity. System x_C^- for glutamate and cystine (Table 1.1) was readily observed in the fetal hepatocyte, but not in the adult hepatocyte after 24 h in primary culture in growth medium (Makowske & Christensen, 1982). This system does reappear, however, when the cells are further cultured for 1 or 2 days in the presence of bovine insulin and dexamethasone (Takada & Bannai, 1984). This result urges us to be cautious in interpreting findings made with the isolated hepatocyte, obviously a cell which is subject to substantial transport modifications. Which observations – those obtained with freshly isolated hepatocytes in suspension, those with hepatocytes obtained in monolayer by primary culture, those obtained by more extended incubation in culture medium, or those made with other more intact preparations – are we to regard as describing the transport potentialities of the hepatocyte in the intact rat? Probably we come nearer to the true picture by taking into account all the transport characteristics detected under various conditions (see also Chapter 5).

Table 1.1 lists a variant Na^+ independent transport system, quite different from the usual system L, appearing along with system L on extended incubation of the hepatocyte after establishing a primary culture monolayer (Weissbach *et al.*, 1982). In contrast, system N appeared to be much the same entity in fetal and mature hepatocytes and in hepatoma cells, in each case being unresponsive to the pancreatic hormones and, as we shall see, responsive both to amino acid abundance and starvation (Vadgama & Christensen, 1983). The Na^+ independent threonine-, alanine-, and serine-preferring system asc of the nucleated red blood cell, and in some phenotypes of its non-nucleated progeny, may prove to contaminate system L transport as observed in various other cells.

1.2.3 *Changes of amino acid transport on fertilisation of oocytes and on activation of the blastocyst*

Epel (1972) showed that a Na^+ dependent transport system appears after fertilization of sea urchin eggs. The finding that valine is transported well by this system, along with other observations, fails however to support assignment of this transport to either system A, or as subsequently proposed (Allemand *et al.*, 1984) to system ASC. Activation of the mouse blastocyst at the end of the diapause leads to the appearance of a broad scope Na^+ dependent transport activity. This system appears unusual since it is inhibited by two bulky bicyclic amino acids, ordinarily largely restricted to Na^+ independent transport systems, and also by its transport of lysine (Van Winkle *et al.*, 1985). A formal demonstration of consistency for the K_i values of this inhibition with the K_m values for uptake showed that these inhibitors are indeed substrates of a shared system, and not merely inhibitors. Furthermore 2^- methylamino isobutyric acid (MeAIB) failed to act as a substrate.

These observations point to a distinct and previously undescribed transport system, characterised by service for branched-chain, bicyclic and even cationic as well as ordinary dipolar amino acids. Its unusual properties, both in invertebrate eggs and the mouse blastocyst, raise the question whether the unusually wide-scope Na^+ independent transport system characteristic of epithelial brush borders, which also accepts the bicycloheptylamino acids (see Ullrich *et al.*, 1974), may represent a reappearance of this system in subsequent ontogeny. The designation system B is proposed, marked with superscript 0, + where desired to indicate the tolerance of net amino acid charge (see Table 1.1).

1.2.4 *Perinatal changes in hepatic amino acid transport: an early observation of adaptive regulation*

Because distinct amino acid transport systems had not been clearly discriminated in 1948, we (Christensen *et al.*, 1948) were not then prepared to learn much from differences among the various amino acids in their competitive interactions (see Figure 1.1). We reported without comment indications of special interaction between glycine and proline for uptake by muscle. When we sought to study the distribution of [^{14}C]-AIB and other model amino acids in the pregnant guinea-pig, we were surprised to note that the fetal liver, compared with the maternal liver, discriminated against these amino acids. Here we were helped by the abnormal structure of our model transport substrate. We observed (Christensen & Clifford, 1963) that the very weak hepatic uptake of these analogues increased abruptly during the first 24 h after delivery of the young (Figure 1.3). During the same interval the hepatic accumulation of glycine increased from five-fold to 20-fold with respect to the plasma level. The gradients thus attained for glycine and AIB approached the values seen in the adult guinea pig. Subsequently we interpreted this change to arise from 'a major intensification of the contribution of a specific transport system' (Christensen, 1973) – undoubtedly system A. For transport system A the repressive factor appears to be the elevation of certain free amino acids, which are maintained in the fetal circulation at a level about five times that applying for the neonatal animal. When the fetal guinea pig is delivered, its tissues as they enter the neonatal state are abruptly subjected to an extracellular environment relatively impoverished in amino acids by the hungrily growing young tissues. These abruptly lowered levels are seen as the de-repressing signal whereby a major intensification of system A transport is evoked.

1.2.5 *Description of adaptive regulation of amino acid transport*

Guidotti *et al.* (1978) commented that the transport response of the guinea pig liver immediately after birth render it an excellent example of a physiological change producing adaptive stimulation of amino acid transport. In the same review the authors wisely recommend the use of cultured cells for observing adaptive regulation, because appropriate physiological stimuli can be

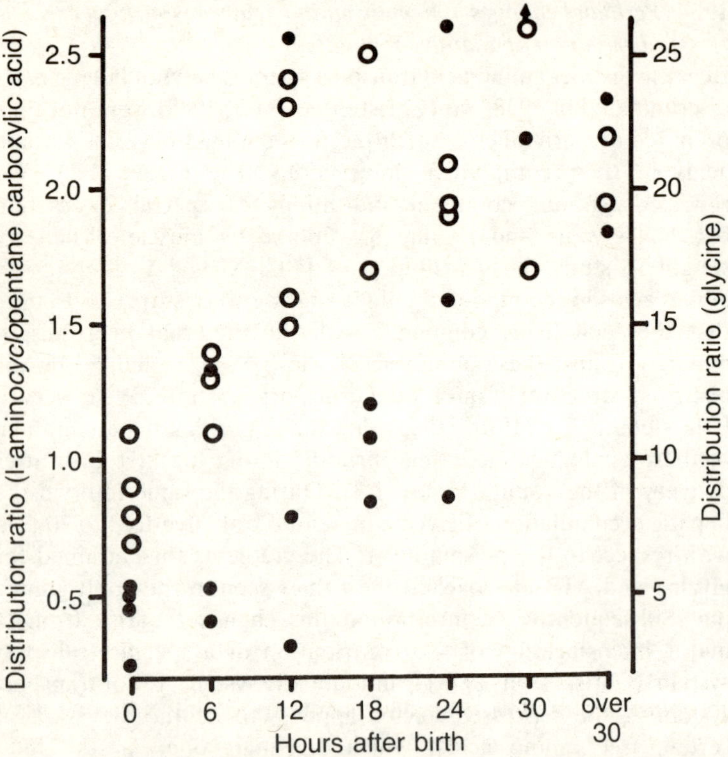

Fig. 1.3 Derepression of transport system A after birth. During the 30 h after birth in the guinea pig changes were observed in the distribution ratios of 1-aminocyclopentane carboxylic acid (o) and glycine (•) between liver cell water and plasma. (Taken with permission from Christensen, H. N. & Clifford, J. B. (1963). *Journal of Biological Chemistry*, **238**, 1743–5.)

applied for extended periods. Riggs & Pan (1972 – see Guidotti *et al.*, 1978) had been able to maintain *in vitro* a separated horn of the weanling rat uterus in artificial medium, initially amino acid-free, long enough (up to 4 h) to observe strongly intensified uptake of AIB, proline, alanine, and cycloleucine. Similar stimulation was observed with estrogen-primed uteri.

The important series of papers by Gazzola, Franchi Gazzola & Guidotti also appeared in 1972 (the same year as Riggs' publication), and their comprehensive review of regulation of amino acid transport in animal cells followed (Guidotti *et al.*, 1978). These studies showed that stimulation of transport specific to system A occurred in embryonic chick heart cells and also in other tissues following amino acid deprivation. Furthermore only some amino acids when present were able to repress this stimulation of transport. Interestingly all these amino acids were substrates of system A. N-methyl-AIB, a non-metabolised analogue specific for system A was shown to be highly repressive – this gives an important indication that formation of a tRNA derivative is not necessary. Leucine, whose rather weak but considerable transport by system A has been demonstrated also appeared to be repressive at 5 mM if a long enough period of contact was allowed (Gazzola *et al.*, 1981).

1.2.6 *How are the signals established for repression and de-repression?*

On the basis of the action of cycloheximide as an inhibitor of protein synthesis and actinomycin D as an inhibitor of RNA synthesis, first with chick heart cells (Guidotti *et al.*, 1978) and then with cultured fibroblasts, Gazzola *et al.*, 1981 concluded that amino acid starvation stimulates system A transport by a de-repression of protein synthesis. This action as described for the fibroblast required a preceding synthesis of mRNA, one that began within 30 min of the change in the extracellular amino acid level, and appeared to continue for 90 min. The transcribed mRNA appeared to persist for a few hours in a state still translatable, presumably into one or more transport proteins, in fibroblasts maintained in the absence of added amino acids. The repressive response appeared also to involve an early mRNA synthesis, whose translation presumably yielded a protein able to cause the degradation or inactivation of transport protein molecules. The decay in transport activity in this situation was somewhat faster in the presence of added system A-reactive amino acids. Gazzola *et al.* (1981) propose a scheme (Figure 1.4) to explain how the concentration of the system A substrates may modulate transcription of mRNA species coding on the one hand for transport proteins, and on the other hand for their presumed inactivators, also how these amino acids may regulate the efficiency of the inactivation of transport proteins. The occupation

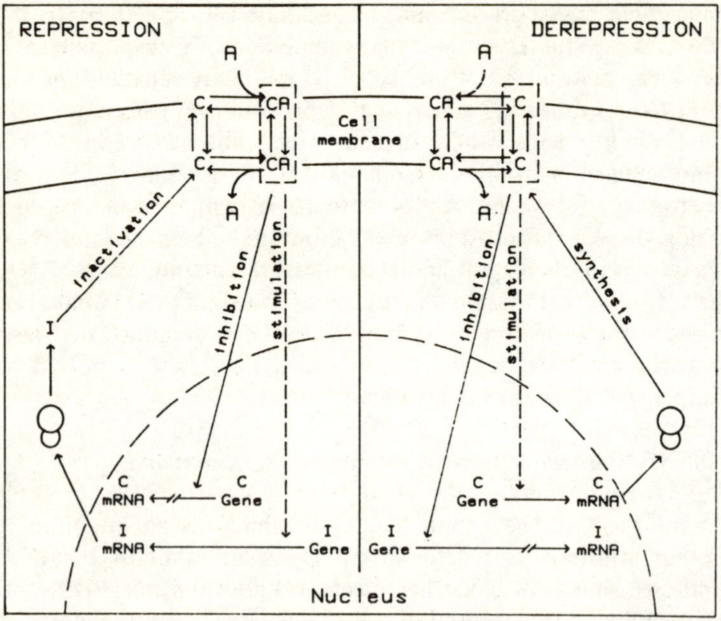

Fig. 1.4 Schematic representation of the mechanisms involved in adaptive regulation of transport System A. A, amino acids reactive with site Al; C, the specific transport protein or 'carrier'; I, the inactivator protein. The dashed-line boxes in the membrane represent the signals for stimulation or inhibition originating from the loaded (CA) or the unloaded (C) transport protein, without any implication as to whether they lie at the inner or the outer surface of the membrane. Note that under conditions of repression the concentration of A is high and under conditions of depression the concentration of A is low. (Taken with permission from Gazzola, G. C. *et al.* (1981). *Journal of Biological Chemistry*, **256**, 3191–8.)

of the transport proteins by the substrate, perhaps in ternary complex with Na^+, is seen as generating a repressive signal inhibiting transcription of mRNA coding for synthesis of the transport proteins, but under conditions allowing and perhaps stimulating transcription of other mRNA coding for the synthesis of an inactivator protein. Occupation of the transport receptor site was therefore held as decisive.

Various investigators have for the most part added confirmation to the phenomenon of adaptive regulation on amino acid starvation, and support for various features described for it. An interesting case was reported by Enders *et al.* (1976) for fragments of the human placenta although as in the case reported by Riggs & Pan (1972 – see Guidotti *et al.*, 1978) no reversible effect of starvation with respect to particular amino acids was established. In the meantime a somewhat similar adaptive regulation has been observed for system N, which transports glutamine, asparagine, and histidine across the plasma membrane of the hepatocyte (Kilberg *et al.*, 1980). Transport by system N of the hepatocyte failed, however, to be repressed by two of its substrates, glutamine and asparagine, although repression was produced by a third substrate, histidine, and also by the non-substrate, MeAIB. These discrepancies from what one might expect from the scheme of Figure 1.4, where the signal for repression is held to flow from the occupied transport receptor complex itself, should be taken seriously, because the mode of transmission of the signal to the gene level (represented by the long downward arrows in Figure 1.4) is unknown. The amino acids whose presence is repressive appear not identical overall with those whose absence can be de-repressive (Bracy *et al.*, 1986).

White & Christensen (1983) were able to manipulate the cellular Na^+ content of cultured hepatoma cells by treating them with ouabain. As a secondary consequence of the changed Na^+ distribution, this agent allowed the balance between influx and efflux of MeAIB to be modulated, thus 'freezing' the cells at a desired degree of transport repression. Ouabain did not directly influence the MeAIB fluxes, and thus gave no kinetic evidence of modifying the occupancy of the system A carrier site. This agent has therefore served to separate the extent of occupation of Site A, on the one hand, from the internal level of MeAIB, on the other, and points to the level of an internal co-repressor as the factor signalling for adaptive regulation of the transport system. The classical hypothesis of Cohen & Monod would tend to interpret a close correspondence between the amino acids transported by a given system and the amino acids able to repress that system to mean that the repressor protein in the chromosome bears a fairly accurate copy of the recognition site for transport. Once that view is accepted, one could quite as well suppose that the copy is only approximate, and resembles the carrier site to different degrees for different transport systems, in the extreme case differing from it to the point where the binding site for a co-repressor is quite different from the transport recognition site.

Moffett & Englesberg (1984) found in Chinese hamster ovary cell lines that β-alanine inhibits de-repression of system A, although it fails to inhibit proline transport. Conversely, they found the unusual system A substrate, L-2, 4-diaminobutyric acid a good inhibitor of system A but a very poor co-repressor of that transport system. These authors have now generated cell mutants, including a mutant deficient in the adaptive ability for system A as measured by MeAIB-inhibitable proline uptake. Instead that system appears permanently de-repressed as to transport (Moffett & Englesberg, 1984). In the proline-negative parent cells, alanine antagonises growth by preventing proline uptake. The mutant cell line is, however, alanine-resistant. This mutant, Ala^r4, is completely resistant to proline-induced inactivation of system A. These authors present a model (Figure 1.5) showing the repressive signal generated at the gene level, with no requirement that co-repressor amino acids be transported by the system modulated. The scheme allows the effective amino acid to act not only to repress the synthesis of mRNA corresponding to the gene coding for a system A transport protein, but also to act as an inactivator, modulating the activity of the transport carrier itself.

White & Christensen (1983) also observed the repressive phase of adaptive regulation to operate somewhat symmetrically on influx of amino acids into and efflux out of the cultured rat hepatoma cell HTC. A more or less symmetrical action of amino acid excess, if it proves generally applicable, does not necessarily render adaptive regulation futile. As long as amino acid transport remains rate-limiting to cellular uptake or release, i.e. as long as metabolism occurs rapidly enough to keep the transport process rate-limiting, a symmetrical modulation of influx and efflux will nevertheless govern the rate of uptake or release (Christensen, 1983). Furthermore it is inherent in the interaction of transport systems that System A will rarely contribute equally to uptake and exodus of an amino acid, as we appreciated long ago from the difference in stereospecificity for these two fluxes in the Ehrlich cell (Oxender, 1965). Evidence that the repressive action on amino acid transport may not be produced by amino acids acting at

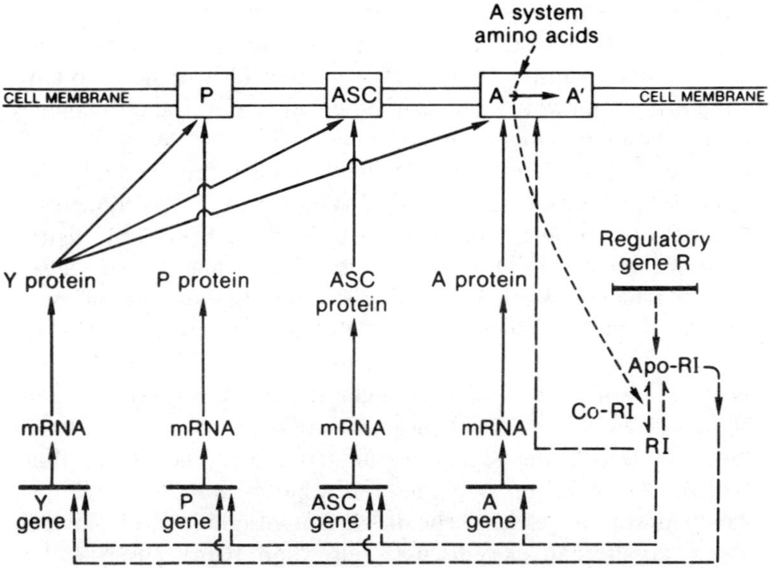

Fig. 1.5 Model for the regulation of the A system of amino acid transport and the regulatory link between the A, ASC, and P systems. See the text for discussion. (Taken with permission from Moffett, J. & Englesberg, E. (1984). *Molecular and Cellular Biology*, **4**, 799–808.)

the recognition site required for the stimulation of transport induced by amino acid starvation has been discussed (Kilberg *et al.*, 1985a).

For the hepatic utilisation of alanine in the rat, transport into the liver has been shown to remain rate-limiting under a wide range of concentrations. Therefore adaptive regulation of System A, as produced by low or high protein intake (Kristensen *et al.*, 1983, Fafournoux *et al.*, 1983; 1985), appears likely to serve as an effective mode of regulation, even if influx and efflux are indeed similarly responsive to it. Regulation of amino acid transport in the hepatocyte *in situ* might become an even more complex subject when the measurements differentiate among the regions of the hepatocyte surface.

System N for glutamine transport across the hepatocyte plasma membrane is judged a potential regulatory site in glutamine degradation and synthesis, given that the movement across the periportal and perivenous surfaces of the cell are considered separately (Haussinger *et al.*, 1985). A tendency to suppose that brush-border transport systems for amino acids will prove to have little in common with systems at other cell surfaces and organelles seems, however, premature.

In general restriction of amino acid supplies and their repletion have been found to influence V_{max} rather than K_m, as though the number of transporter molecules were increased and decreased in adaptive regulation. For the myogenic cell line L6, however, amino acid starvation changed the properties of System A as measured by Na^+ dependent proline uptake with regard to pH optimum and sensitivity to N-ethylmaleimide inhibition (Klip *et al.*, 1982). Changes of this sort might mean that transport by more than a single system was being measured, one component being more subject to the regulatory influence than the other; alternatively they could be accounted for by kinetic complexity within a single transport system. The important discrimination between these two alternatives is discussed elsewhere (Christensen, 1985). Kinetic evidence for an as yet undefined new complexity for MeAIB transport is illustrated in Figure 1.6. With either transformation or amino acid deprivation, Boerner & Saier (1985) reported these findings for the kidney-derived MDCK cell line in monolayer culture. The laboratories of the present authors have both observed provisional evidence for such dichotomies of System A transport in various types of cell.

In the cultured fetal human fibroblast (Gazzola *et al.*, 1984), although not in the hepatocyte and certain other cells (see Kilberg *et al.*, 1985b), blood serum needs to be present for the full expression of the enhancement of System A transport by amino acid deprivation. The serum factors involved have not been identified, but for the fetal fibroblast, insulin addition could not substitute for serum. Figure 1.7 uses proline uptake to measure the stimulation of System A transport in the human fibroblast. The large stimulation produced by several hours of amino acid deprivation in the presence of serum, or the much smaller stimulation obtained in the absence of serum was not obtained if 18 μM cycloheximide was present, and hence this stimulation depends on macromolecular synthesis. If the serum is added later, during the course of amino acid deprivation rather than at first, the

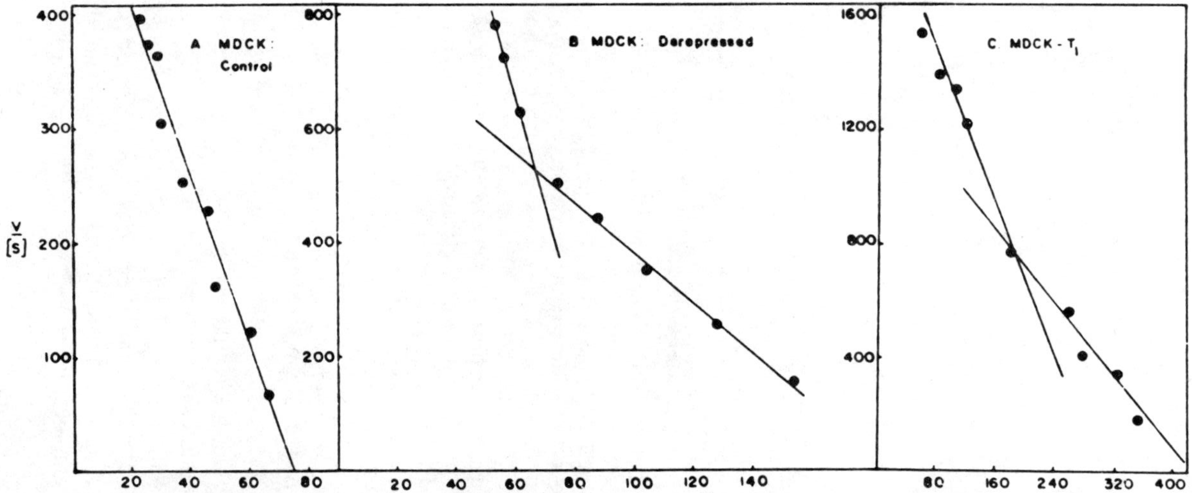

Fig. 1.6 *Kinetic* study of regulation of System A. Eadie-Hofstee plots of Na^+ dependent MeAIB uptake (v in pmol · min^{-1} · mg^{-1} protein) as a function of the MeAIB concentration (mM), for amino acid-fed (Panel A) and for amino acid-starved (Panel B) MDCK cells, or for a transformed variant (Panel C) of this cell line. See text for discussion. (Taken with permission from Boener, P. & Saier, M. H. (1985). *Journal of Cellular Physiology*, **122**, 308–15.)

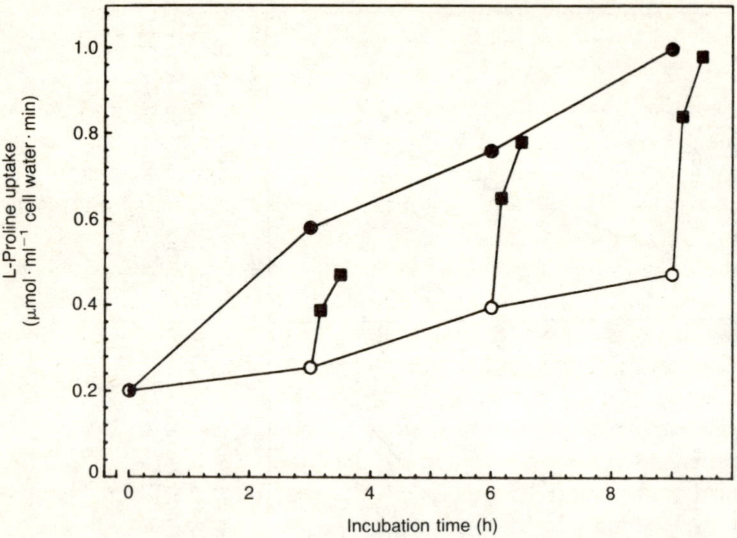

Fig. 1.7 Evidence for separate synthesis and membrane insertion of system A carriers. Changes in L-proline uptake were used to measure system A activity by cultured fetal human fibroblasts incubated in amino acid-free medium. The effect of adding serum at various time intervals during amino acid starvation is shown. Cells were incubated for 9 h in balanced salt solution at 37°C in the absence (0) or in the presence (•) of 15% fetal bovine serum. To produce the curves marked ■, the serum was added to cells incubated for 3, 6, and 9 h in the serum-free medium (in each case supplemented with cycloheximide 15 min before serum addition). Incubation was continued 20 min after the serum addition. At the times indicated, cell monolayers were washed and tested for the rate of [^{3}H]-L-proline uptake during a one minute incubation. (Taken with permission from Gazzola *et al.* (1984). *Biochemical and Biophysical Research Communications*, **120**, 172–8.)

prior deficit in stimulation due to the absence of serum is largely made up during a few minutes (steep curves in Figure 1.7) whether cycloheximide is present or not. This change occurs largely through increases in V_{max} rather than alterations in K_m. The effect of serum addition was roughly proportional to the amount of transport protein presumably generated within the cell by the de-repressive stimulus.

To quote Gazzola *et al.* (1984):

These results can be interpreted to indicate that serum is essential for a vectorial post-translational event leading to the insertion of transport proteins into the cell membrane.

Evidence had earlier been obtained for the recruiting of membrane protein molecules from intracellular stores for service for glucose transport into adipocytes on insulin stimulation and recent work by Oka & Czech (see Czech, 1985) using photolytic marking with cytochalasin B of such molecules when exposed on the hepatocyte surface has supported this concept (see papers by Simpson & Cushman and by Kono in Czech, 1985). A striking feature of the post-translational stimulation of amino acid transport, given that time has previously been allowed for stimulated macromolecular synthesis during amino acid starvation, is the relative quickness of this effect. The distinction between the processes responsible for the synthesis and the insertion of amino acid transporters by the speed at which these two events occur may well be important in studying the complex mechanisms of transport regulation. Additionally this distinction may help answer the associated question as to why some cells need one or more serum factors for full adaptive regulation, and others do not. These results might suggest that the serum factors needed for the full post-translational component of the adaptive regulation make their own humoral contribution to the regulation of amino acid transport.

A different Na^+ independent transport system particularly vigorous in the transport of cysteine and glutamate in the human fibroblast, and designated x_C^- (see Christensen, 1984; see also Table 1.1) was shown to be stimulated by starvation of cystine or glycine plus serine (Bannai & Kitamura, 1982 – see Bannai, 1984; Dall'Asta *et al.*, 1983). The adaptation is time-dependent and requires synthesis of RNA and protein. Even though glycine is not a substrate of this system, the response to glycine and serine starvation with respect to either cystine or glycine tends to deplete the intracellular glutathione, the latter change then producing the stimulation (Bannai & Kitamura, 1982 – see Bannai 1984a). Treatment of fibroblasts with electrophilic reagents that cause the consumption of cellular glutathione also stimulate transport by system x_C^- (Bannai, 1984b; Bannai *et al.*, 1984). Although hepatoma cell lines show system x_C^- transport activity, isolated hepatocytes from adult rats show little such activity (Makowske & Christensen, 1982). The activity emerges, however, on holding the cells in primary culture in the presence of insulin and dexamethasone for one or two days (Takada & Bannai, 1984).

1.2.7 *Adaptive regulation of system L*

Given that transport system L can serve more importantly for cell release than for cell entry of several of the amino acids (see Figure 1.2), one should not be surprised to find a different basis for its physiological regulation than for systems A or ASC, a difference which recent studies seem to confirm. Early work showed that the response of system L in several cell types to general amino acid starvation was a slight decrease in activity (Moore *et al.*, 1977; Shotwell *et al.*, 1983). The failure to show de-repression of system L transport was attributed by Oxender and his associates to an inability to lower sufficiently the intracellular levels of one or more regulatory substrates of system L by direct leucine starvation, provided that other nutrients are not restricted.

By the use of dialysed fetal bovine serum, the leucine concentration in the Chinese hamster ovary (CHO) cell line has been brought low enough to show a de-repressive effect. A stronger effect has been obtained, however, in a temperature-sensitive CHO mutant. When grown at high temperatures, these cells become starved of leucine because of their limited ability to form leucyl-tRNA, and, presumably as a response, show increased System L transport activity. This increase, which can be prevented by adding high levels of leucine, can thus be seen in this mutant line (Moore *et al.*, 1977; Shotwell *et al.*, 1982, see Shotwell *et al.*, 1983). Phenotypic revertants of this mutant have been isolated which have increased System L transport activity, and are defective in their regulation of this system. In these tests the regulated transport activity was shown to be restricted to Na^+ independent transport and was inhibited by the relatively system L-specific bicycloheptane analog (see Table 1.1). The enhancement could be prevented by cycloheximide but not by actinomycin D. Accordingly regulation at the level of translation is suggested, determined by the occupancy of the corresponding tRNA. As already mentioned, non-mutant CHO cells also showed stimulation of system L at very low levels of leucine, providing other nutrients were not restricted. Hence a significant difference arises here between adaptive regulation of systems A and L. Although several artificial substrates largely restricted to system L are known, none of these has apparently so far mimicked the regulatory action seen here with leucine alone. It is not clear how this mode of regulation of system L serves in the intact animal at physiological leucine levels. During incubation of rat hepatocyte monolayers, adaptive regulation has also been observed for a particularly high affinity, Na^+ independent component of amino acid transport designated system L1 (Weissbach *et al.*, 1982). The physiological regulation of this component is not yet well understood. This component as already mentioned is associated with somewhat the same group of amino acids transported by the classical system L, but can be distinguished from that system on the basis of substrate selectivity as well as by its high amino acid affinity. In general, there seems to be much to be learned yet before we fully understand how the regulation of the several transport systems produces the needed coordination of interorgan and intracellular amino acid flows.

1.3 The present status of the endocrine regulation of amino acid transport

1.3.1 *Physiological background*

Despite the existence of adaptive regulation at the cellular level, the multi-organ animal requires additional levels of control. Division of labour among the tissues of the body is a fundamental concept in biology; yet, such diversification requires a sophisticated coordination scheme so that the proper response occurs to both intracellular and extracellular stimuli. Hormonal regulation of nutrient flows among the various organs represents an important process by which animals maintain this coordination.

Hormone effects on amino acid transport have been reviewed previously for a variety of tissues (Riggs, 1970; Guidotti *et al.*, 1978; Shotwell *et al.*, 1983) and are summarised here in Table 1.2. In most instances, System A-mediated amino acid uptake is stimulated by a particular hormone if that hormone causes an increase in the anabolic state of the tissue. Conversely, transport

Table 1.2 *Regulation of amino acid transport by hormones and growth factors*

Effector Agent	Effect on amino acid transport	Tissue
Hormones		
ACTH	increased	adrenal gland, muscle, intestine, kidney
	decreased	adipose
Androgens	increased	prostate gland, seminal vesicles, skeletal muscle, kidney
Catecholamines	increased	liver, kidney, heart, skeletal muscle, intestine
	decreased	adipose, skeletal muscle
Estrogens	increased	uterus, Ehrlich ascites cells
FSH	increased	ovaries, testes
Glucagon	increased	liver
	decreased	adipose
Glucocorticoids	increased	liver, fibroblasts
	decreased	skeletal muscle, thymocytes, lymphocytes, mammary gland, HeLa cells, hepatoma cells, liver, diaphragm
Growth hormone	increased	skeletal muscle, heart, liver, adipose, diaphragm
Insulin	increased	skeletal muscle, diaphragm, bone, heart, liver, adrenal gland, lymphocytes, uterus, adipocytes, fibroblasts, hepatomas, mammary gland, exocrine pancreas, thyroid gland, retina, lung
LH	increased	ovaries
PTH	increased	bone, calvaria, kidney cortex
	decreased	renal resorption
Prolactin	increased	intestine, mammary gland, skeletal muscle, thyroid gland, liver
Thyroid hormones	increased	bone, cartilage, thymocytes, liver, intestine, pituitary
Growth factors		
EGF	increased	chick embryo heart cells, NRK-49F cells
	decreased	liver

Table 1.2 (*continued*)

Effector Agent	Effect on amino acid transport	Tissue
FGF	increased	fibroblasts
IGF-II (MSA)	increased	skeletal muscle, hepatoma, chick embryo heart, liver
IGF-I (Somatomedin C)	increased	human fibroblasts
NGF	increased	phaeochromocytoma cells
TGF	increased	NRK-49F cells, rat fibroblasts

ACTH = adrenocorticotrophic hormone;
FSH = follicle stimulating hormone;
LH = lutenising hormone;
PTH = parathyroid hormone;
EGH = epidermal growth factor;
FGF = fibroblast growth factor;
IGFII (MSA) = insulin-like growth factor II;
NGF = nerve growth factor;
TGF = thymocyte growth factor.

is either unaffected or inhibited if the hormone's net effect is catabolic in nature. The development of culturing procedures for isolated cells and tissues has allowed investigators to determine the primary effects of hormones on membrane transport. Furthermore, the availability of transformed cells following oncogene transfection has allowed Racker and his associates to initiate a study on the effect of transforming growth factors on amino acid transport and metabolism in tumour cells (Racker, *et al.*, 1985).

In the context of amino acid metabolism, hormonal control can be illustrated by the sequence of events that occur after ingestion of a high protein meal. The elevated levels of plasma amino acids cause an increase in the release of the pancreatic hormones insulin and glucagon. Either insulin (Kletzien *et al.*, 1976; LeCam & Freychet, 1978) or glucagon (LeCam & Freychet, 1976; Pariza *et al.*, 1976a), or both, may then act on the liver to increase amino acid uptake via System A. The resultant accumulation of amino acids may be followed by, or coupled to, insulin-stimulated protein synthesis or glucagon-induced gluconeogenesis.

1.3.2 *The hepatocyte as a model*

Hormonal control of amino acid transport in liver has been investigated extensively, perhaps to a greater extent than in any other tissue. Hence, rat liver will be used as a model to outline the current understanding of hormonal regulation of transport. Three observations implicate System A as the primary transport process that is capable of responding to hormones. First, the transport System affected by hormones is Na^+ dependent (LeCam & Freychet, 1976), thus ruling out mediation by System L. Second, the uptake of MeAIB, a System A-specific analogue, can be used to monitor the hormone-sensitive System (Kilberg *et al.*, 1980); a result shown for a few other cell types as well (Guidotti *et al.*, 1974; Wheeler *et al.*, 1980, in Kilberg *et al.*, 1985a). Third, the hepatic Systems ASC, Gly, and L are unaffected by hormone action when measured directly using the relatively System-specific substrates cysteine (System ASC), glutamine or histidine (N), glycine (Gly), or BCH (L).

1.3.3 *Hormones which influence transport*

The effect of hormones on the kinetics of amino acid uptake has been reported for a large number of tissues and cell types. In general, the results show that hormones alter the V_{max} of the transport system, rather than the K_m for the substrate amino acid (Shotwell *et al.*, 1983). Caution must be exercised when interpreting transport data based on the kinetic parameters, K_m and V_{max} (Christensen, 1975). It is generally assumed that an increase in the V_{max} of a transport System is the result of an increase in the number of active carriers within the membrane. However, the K_m for a transport substrate should not be taken as a direct measure of the dissociation constant of the substrate-carrier complex (Christensen, 1975). Furthermore, translocation of an amino acid by multiple transport Systems necessitates additional caution in assignment of K_m values.

Amino acid transport via System A appears unique among the hormonesensitive processes of the cell because of its ability to respond to a wide variety of hormones. Examination of Table 1.2

will show that more than 20 different hormones or growth factors produce changes in amino acid uptake; nearly every tissue of the body that has been tested exhibits hormonesensitive transport. Attempts to propose a model for hormone-mediated induction of System A are complicated by the knowledge that some of these hormones act through plasma membrane receptors, whereas others interact with nuclear receptors. Clearly, this spectrum of molecules with transport-stimulating capability must be reflected in any hypothesis for hormonal control of amino acid uptake. Our current knowledge of cell function would lead us to suppose that there are multiple pathways by which the hormone-mediated stimulation of transport occurs. Two possible areas of interest will be discussed below: *de novo* synthesis of a protein necessary for transport activity; and post-translational conversion of transport carriers from an inactive to an active state.

1.3.4 *Mechanism of hormonal action*

Numerous investigations have focused on the role of *de novo* protein synthesis in the stimulation of amino acid transport. Most of these investigations have utilised protein synthesis inhibitors such as puromycin or cycloheximide, or compounds that block RNA biosynthesis such as actinomycin. The data shown for glucagon dependent (Figure 1.8A) or insulin-dependent (Figure 1.8B) stimulation of Na^+ dependent AIB uptake (i.e., System A activity) in cultured rat hepatocytes are typical of the results reported by many laboratories for a variety of hormones. The data given for glucagon demonstrate that there is a lag of about an hour before the hormone effects become apparent (Figure 1.8A). During the next several hours, the glucagon causes a continuous increase in the transport activity. It should be noted that the stimulation of System A occurs within the normal physiological range for both insulin and glucagon. Interestingly, System A represents the only extensively characterized cellular process for which insulin and glucagon fail to show antagonistic actions (LeCam & Freychet, 1978). In the presence of cycloheximide or actinomycin, the hormone-induced response is completely abolished (Figure 1.9a). Similar results can be obtained with alternative inhibitors such as puromycin for protein synthesis, α-amanitin for total RNA synthesis, and cordycepin or ade-

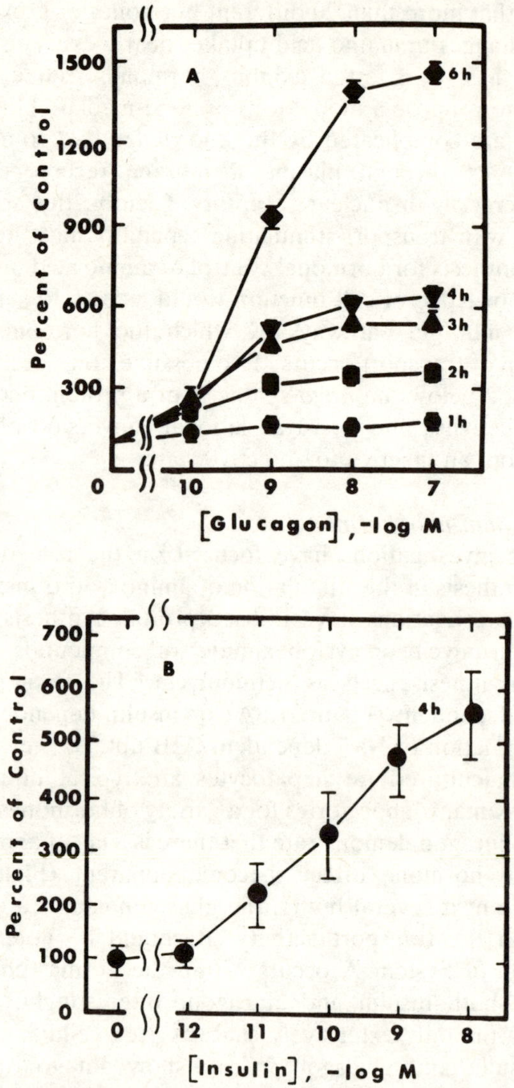

Fig. 1.8 Induction of hepatic system A transport activity by either glucagon or insulin. Freshly isolated hepatocytes were suspended in Waymouth's medium supplemented with hormone at the concentrations indicated. The cells were incubated with glucagon for 1, 2, 3, 4, or 6 h (A) or with insulin for 4 h (B) prior to assaying the System A activity. The Na^+ dependent uptake of 50 μM AIB was measured for 1 min at 37°C. The data are average ±SD, of at least three individuual assays. The control value (100%) was 16.4 ± 3.2 pmol · mg^{-1} protein · min^{-1}.

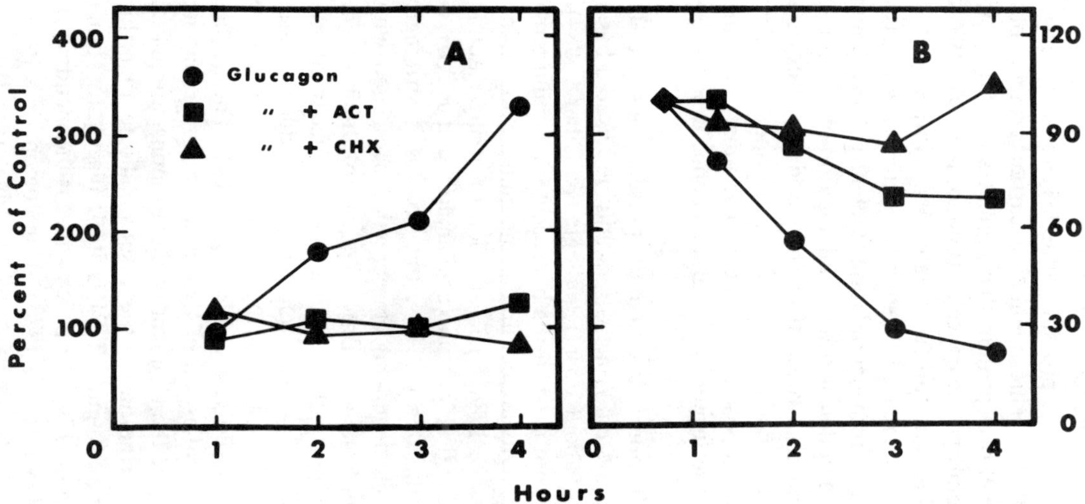

Fig. 1.9 Inhibition of the induction and decay of the glucagon-induced hepatic system A activity. (A) Normal hepatocytes in primary culture were incubated in the presence of Krebs-Ringer bicarbonate buffer containing 10^{-7} M glucagon (●), glucagon plus 4 μM actinomycin (■), or glucagon plus 100 μM cycloheximide (▲). System A activity was monitored by assaying Na^+ dependent AIB (50 μM) uptake for 1 min at 37°C. The control value was 10.8 pmol · mg^{-1} protein · min^{-1}. (B) Hepatocytes were isolated from rats that had been injected with glucagon (1 mg per 100 g body weight) 4 h prior to surgery. The cells were suspended in Waymouth's culture medium and placed in primary culture (●). In some instances, the culture medium contained 4 μM actinomycin (■) or 100 μM cycloheximide (▲). After the indicated interval of culture, System A activity was assayed as described above. The control value was 314 pmol · mg^{-1} protein · min^{-1}. For both experiments, the standard deviations for triplicate determinations were typically less than 10%.

nine-9-β-D-arabinopyranoside (ara-A), for poly(A)-containing mRNA biosynthesis (Handlogten *et al.*, 1986).

It is possible that the role of protein synthesis is not to generate increased amounts of a specific membrane carrier protein, but rather to alter the level of a protein that regulates the transport activity. For chick embryo heart cells (Guidotti *et al.*, 1978) concluded that insulin not only stimulates the synthesis of a protein necessary for increased System A activity, but also protects the transporter from degradation. Gelehrter (1979) has proposed that the dexamethasone-induced inhibition of System A in HTC hepatoma cells may result from increased synthesis of a protein involved in the degradation of the transport System. Similarly, it has been proposed that glucagon stimulates hepatic System A by slowing the turnover rate of a component necessary for transport activity (Pariza *et al.*, 1976a, b).

In the absence of the hormone and in the presence of substrate amino acids, elevated System A activity rapidly decays to a basal level with a half-time of about 1.5 h (Handlogten & Kilberg, 1984). This inactivation does not occur if the hepatocytes containing the glucagon-stimulated activity are cultured in medium lacking amino acids (Handlogten *et al.*, 1986). The addition of a single amino acid, even a non-metabolisable amino acid analog such as AIB, is sufficient to initiate the decay. Thus, this process is strikingly similar to the amino acid dependent repression associated with adaptive regulation for System A (Bracy *et al.*, 1986). Inhibitors of either RNA or protein synthesis can block the decay of the enhanced transport (Figure 1.9); even if they are added several hours after the decay has begun, further loss of transport activity will be prevented (Handlogten & Kilberg, 1984). These data suggest that a newly synthesised mRNA and the corresponding protein are necessary for the turnover of the hormone-stimulated System A activity. Furthermore, the results indicate that this 'Transport-Inactivating-Protein' (TIP) is itself inactivated quite rapidly ($t_{1/2}$ = 1 h), probably by a process that is independent of protein synthesis (Pariza *et al.*, 1976a; Handlogten & Kilberg, 1984). Interestingly, inhibitors of poly(A) polymerase such as cordycepin (Handlogten *et al.*, 1986) or ara-A (Figure 1.10) do not block the amino acid-dependent repression of glucagon-induced System A activity.

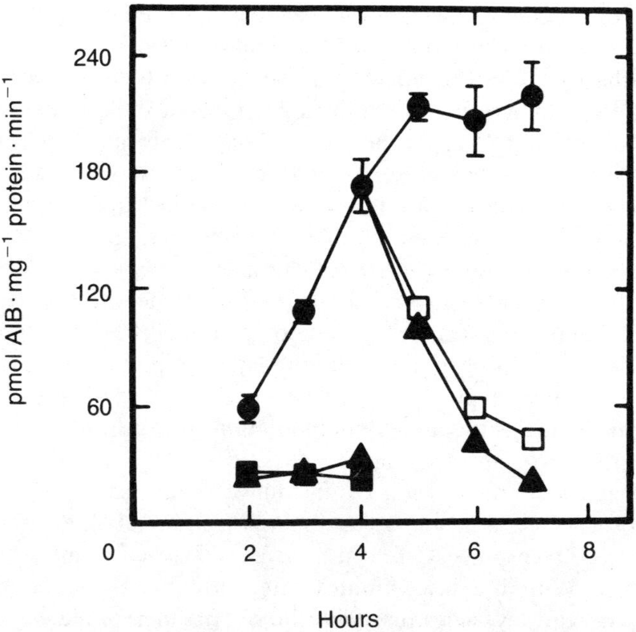

Fig. 1.10 Effect of adenine-9-β-D-arabinopyranoside (ara-A) on adaptive regulation in cultured rat hepatocytes. Cells in primary culture were transferred to (●) amino acid-free (AAF) medium; (■) AAF medium + 100 μM ara-A; (▲) amino acid-supplemented (AAS) medium; (□) AAS medium + 100 μM ara-A. The starvation dependent de-repression and amino acid-induced repression of system A activity were monitored by assaying Na^+-dependent AIB (50 μM) uptake for 1 min at 37°C. The results were similar when either Krebs-Ringer bicarbonate buffer + 10 mM asparagine or Waymouth's culture medium was used as the AAS medium. The data are average ±SD, of four determinations. Where not shown the standard deviation bars are contained within the symbol.

In a few tissues, the stimulation of amino acid transport has been reported to be independent of protein synthesis; an observation illustrated either by an insensitivity to inhibitors or by induction so rapid as to exclude the possibility of *de novo* protein synthesis. Examples include: the insulin effects in mammary tissue (Lobitz & Neville, 1977), diaphragm (Wool *et al.*, 1965; Kostyo, 1968) and frog skeletal muscle, (Grinstein, & Erlij, 1977); growth hormone enhancement in diaphragm (Knobil, 1966).

Several possible mechanisms can be proposed to account for hormonal control by routes independent of protein synthesis. One hypothesis is the conversion of the transport carrier from an inactive to an active state. An inactive form of the carrier may be synthesised and inserted into the plasma membrane. Obviously, the initial step in this process would require protein synthesis, but it would not result in a detectable increase in transport activity. The hormone could then stimulate nutrient transport by causing an increase in the conversion of the inactive carrier to its active form by a mechanism independent of protein synthesis (Pariza *et al.*, 1976b). Post-translational modification of the carrier protein represents one process by which the latter step might occur.

An alternative possibility by which the cell might achieve a protein synthesis independent stimulation of transport is through recruitment of intracellular carriers. This model can best be illustrated by the effect of insulin on glucose transport in adipocytes (see Simpson & Cushman in Czech, 1985). An 'excess' of glucose transporters are synthesised and stored in intracellular vesicles. Within a few minutes after insulin has bound to its receptor, these vesicles fuse with the plasma membrane and thus, cause a rapid, protein synthesis independent stimulation of glucose transport. After removal of the hormone, the transporters are recycled by a reversal of this exocytotic process. Some cells may have the potential to store amino acid transport carriers in much the same way such that one might detect stimulation of uptake that is insensitive to inhibitors of RNA or protein synthesis. Indeed, as discussed above, the serum effects on adaptive regulation of System A in human fibroblasts (Gazzola *et al.*, 1984) and the amino acid dependent induction of System N in cultured hepatocytes (Weissbach & Kilberg, 1984) are compatible with such a proposal.

1.3.5 *Unanswered questions on hormonal regulation*

In summary, hormonal regulation of amino acid transport at the plasma membrane represents a fundamental response of the cell to its environment. Although the process is not well understood at the molecular level, a considerable amount of information has been published using intact cells. Most of the reports dealing with hormonal control of amino acid transport have been descriptive in nature, outlining changes in kinetic parameters, relating hormone binding to enhancement of transport, and describing similar characteristics. The number and sequence of events that occur between hormone binding and the observed stimulation of transport is, for the most part, unknown. In fact, even the cellular location of the protein responsible for stimulated System A activity remains to be determined.

The following are included among the many questions yet to be answered:

(i) Is there direct interaction between membrane proteins involved in hormone binding and those responsible for transport?
(ii) How is the rate of transcription or translation for the System A-associated protein controlled at the molecular level?
(iii) Does selective protein turnover, either within the plasma membrane or in the cytoplasm, play a role in membrane transport regulation?

Research aimed at answering these questions will provide a beginning in our efforts to understand, at the molecular level, how hormones enhance amino acid transport. Clearly, our present lack of knowledge should be regarded as a continuing challenge to those already in the field and as an area of great research potential for those new to the subject.

1.4 **Further reading**

(Other literature is listed in the compiled References at the end of the book).

Christensen, H. N. (1975). *Biological transport*, 2nd Edn. W. A. Benjamin, Reading, Mass.

Christensen, H. N. (1982). Interorgan amino acid nutrition, *Physiological Reviews*, **62**, 1193–1233.

Christensen, H. N. (1984). Organic ion transport during seven decades. The amino acids. *Biochimica et Biophysica Acta*, **779**, 255–69.

Christensen, H. N. (1985). On the strategy of kinetic discrimination of amino acid transport systems. *Journal of Membrane Biology*, **84**, 97–103.

Christensen, H. N. *et al*. (1986). Interorgan Transport. *Federation Proceedings*, **45**, 2165–83.

Guidotti, G. G., Borghetti, A. F. & Gazzola, G. G. (1978). The regulation of amino acid transport in animals cells. *Biochimica et Biophysica Acta*, **515**, 329–66.

Kilberg, M. S., Barber, E. F. & Handlogten, M. E. (1985). Characteristics and hormonal regulation of amino acid transport system A in isolated rat hepatocytes. *Current Topics in Cellular Regulation*, **25**, 133–61.

Kilberg, M. S. (1986). System A-mediated amino acid transport: metabolic control at the plasma membrane. *Trends in Biochemical Science* 11, 183–86.

Oxender, D. L. & Christensen, H. N. (1963). Distinct mediating systems for the transport of neutral amino acids by the Ehrlich cell. *Journal of Biological Chemistry*, **238**, 3686–99.

Shotwell, M. A., Kilberg, M. S. & Oxender, D. L. (1983). The regulation of neutral amino acid transport in mammalian cells. *Biochemica et Biophysica Acta*, **737**, 267–84.

2

The sodium gradient hypothesis of organic solute transport with special reference to amino acids

A. A. Eddy Department of Biochemistry and Applied Molecular Biology, University of Manchester Institute of Science and Technology.

2.1 Concentration gradients of organic solutes formed by epithelia and other tissues

Sacs of everted small intestine, fed with oxygen in an appropriate Ringer solution, concentrate certain carbohydrates and amino acids, during their passage through the tissue from the solution bathing its mucosal face to the corresponding serosal solution (Figure 2.1*a*; Table 2.1). This phenomenon, in which sodium ions and ATP are intimately involved, provides some of the best studied examples of transcellular active transport underlying the absorption of nutrients from the gut into the vascular circulation. Similar mechanisms appear to be involved in the Na^+ dependent reabsorption of carbohydrates or amino acids from the glomerular filtrate across the wall of the proximal tubular epithelium. In both tissues solute absorption, accompanied by the movement of water and electrolytes, takes place across the brush border epithelium, the solute initially being concentrated within the epithelial cells. These occur as a convoluted sheet of cells of characteristic morphology with a surface area enlarged by numerous projections, the microvilli. The coherence of the epithelium is maintained by specialised junctions between the cells themselves (Figure 2.1*b*) comprising pores that tend to exclude amino acids or carbohydrates while permitting a controlled movement of water and simple electrolytes. The epithelium thus exhibits a finite resistance to the passage of an electric current, indeed it becomes electrically polarised during the transport of nutrients.

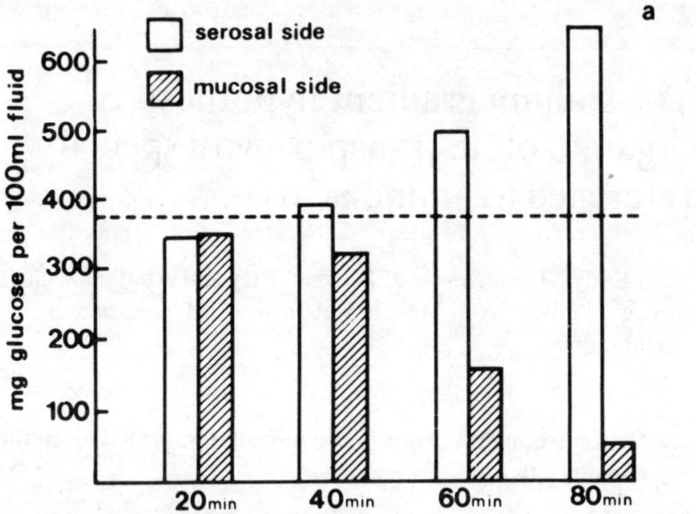

Fig. 2.1(*a*) Time course of the glucose concentration gradients developed by sacs of everted rat intestine. The sac initially contained the same glucose concentration on both the mucosal and serosal sides. (Taken with permission from Wilson, T. H. & Wiseman, G. (1954). *Journal of Physiology*, **123**, 116–25.)

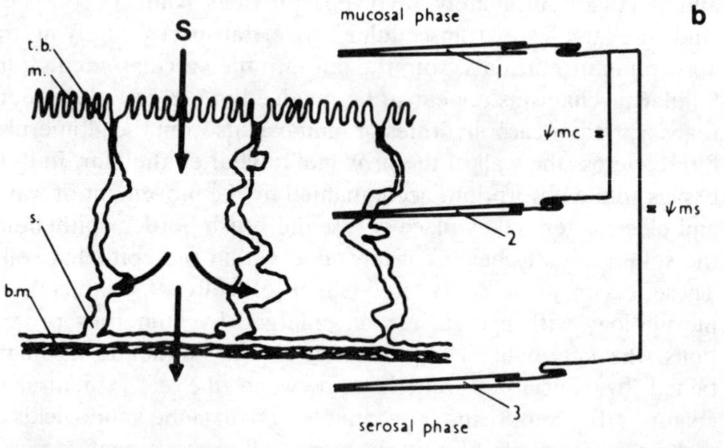

(*b*) Electrical characteristics of the intestine in relation to transfer of solute S. A sheet of absorptive cells is shown, illustrating the mucosal face with villi (m), the spaces (s) between the cells and their basement membrane (b.m.). Terminal bars (t.b.) between the cells increase the resistance to current flow through that pathway. Three electrodes (1,2,3) record the voltage between the mucosal solution and cell (ψ_{mc}) or the voltage across the tissue (ψ_{ms}).

When the tissue expends metabolic energy to move the solute from a solution in which its electrochemical potential is μ_1 to one in which its electrochemical potential is μ_2, the free energy increase involved in transferring one mole of solute S, is

$$\mu_2 - \mu_1 = RT \ln a_2/a_1 + z\mathrm{F}(E_2 - E_1) \qquad (2.1)$$

where R is the gas constant, T the absolute temperature, F the Faraday and z is the formal ionic charge carried by the species S. $E_2 - E_1$ is the difference in electrical potential between solutions 1 and 2, whereas a_1 and a_2 are the activities of S in the respective solutions. Some examples are listed in Table 2.1.

Eqn. (2.1) provides a **minimum** estimate of the amount of energy that the tissue expends per mole of S transferred from solution 1 to solution 2. Because the mechanism of energy coupling is unlikely to be perfectly efficient, more metabolic energy will need to be expended for that purpose than the

Table 2.1 *Typical magnitudes of the concentration gradients of amino acids or carbohydrates formed in various systems*

System	$[S]_i/[S]_0$	$RT\ln([S]_i/[S]_0)$ (kJ · mol^{-1})
α-Methylglucoside in pigeon enterocytes	70	10.9
α-Methylglucoside in pig kidney cells[a]	550	16.9
L-Alanine in hepatocytes	20	7.7
Glycine in pigeon erythrocytes	30	8.8
D-Galactose in bullfrog intestinal cells	3.9	3.5
D-Galactose in hamster intestine	18	7.4
2-Aminoisobutyrate in mouse ascites tumour cells	45	9.8

[a]From Moran *et al.*, 1983 (see Table 2.3 for other relevant literature).

With the exception of the observations with galactose, the maximum concentration gradient the preparation formed from relatively dilute solutions is reported. The free energy change corresponding to the hydrolysis of ATP, the likely source of metabolic energy, is about 47 kJ · mol^{-1} under physiological conditions, representing a factor of about 10^8-fold.

equation indicates. A further possible consequence of such inefficiency is that once the distribution of S between solution 1 and solution 2 has reached a steady-state, at a certain maximum value of the ratio $[S]_2/[S]_1$, a continuous expenditure of metabolic energy might be needed in order to maintain that steady state. This notion that 'slippage', as it were, might occur between the energy input into the solute pump and the energy output represented by the solute gradient is given a more specific meaning in the mechanisms discussed in section 2.4.

2.2 Carrier-mediated solute flow and counter-flow of two solutes competing for the same binding site: the monovalent carrier catalysing facilitated diffusion

A model which has proved widely applicable in discussions of the mechanism of solute uptake and which can be analysed by the mathematical procedures used by enzyme kineticists is shown in Figure 2.2*e*. As a result of thermal agitation, the carrier, C, alternately presents a single binding site towards either the outer or the inner solutions that the membrane separates. When the binding site is accessible to the solute S then CS may be formed. Translocation of CS may lead to subsequent release of S on the opposite side of the membrane. The translocation of carrier, C, or CS is thus associated with a complex process in which the conformation of the protein around a binding site changes so that access of the solute to it is restricted to alternate faces of the membrane.* That the inwards and outwards facing conformations really exist is suggested by the way certain inhibitors interact preferentially with one of the two carrier conformations (Appleman & Lienhard, 1985).

Carriers exhibit the so-called counterflow phenomenon which arises when one solute, A, is present at equal concentrations on both the cis and trans sides of the membrane, a second solute, B, of the carrier mechanism then being added to the cis solution.

---

*Lauger (1980) has argued persuasively that a mobile carrier might be preferentially rather than exclusively accessible from the cis or trans side in its alternative configurations. It would then show some of the characteristics of a channel or pore.

Here it competes for entry with A by forming CB, thereby retarding the inflow of A relative to its outflow. The spontaneous inflow of B down its concentration gradients thus drives an uphill movement of A.

2.3 The sodium gradient hypothesis: solute flow coupled to an ionic gradient on a bivalent carrier

The possibility that a flow of alkali cations (K^+) through a carrier mechanism might propel a counterflow of glycine into mouse ascites tumour cells was first suggested by Christensen and his colleagues as an extension of the counterflow principle (Riggs *et al.*, 1958). In the terminology proposed by Mitchell (1963), such a model represents an **antiport** relationship between the two flows (Figure 2.2*b*), and implies that the carrier binds the respective ligands at two distinct sites. Work with the intestine, by Crane (1962) and others, subsequently focussed attention on the central role of Na^+ and the possibility that the inwards flow of Na^+ might propel a parallel flow of the sugar into the intestinal cell (Schultz & Curran, 1970). Such a model envisages a **symport** relationship (Figure 2.2*a*) between Na^+ and the sugar. the energy inherent in the gradient of Na^+ being used to drive solute transfer across the mucosal boundary by a purely physical process. The requisite sodium gradient is set up in the first place by the sodium pump which thus serves as the primary transducer of chemical energy into osmotic work. In this context the most likely location of the sodium pump, at the basolateral surfaces (Schultz & Zalusky, 1964), was that envisaged by Ussing in a classical model of Na^+ transfer across epithelia (Figure 2.3*a*).

The above work served to focus attention on three important questions bearing on the validity of the sodium gradient hypothesis.

(i) Which inorganic ions interacted directly with the carrier mechanism transferring a given amino acid or carbohydrate across the cell membrane, and which of them were co-substrates?

(ii) Would the presence of an appropriate ionic gradient of the co-substrate acting across the cell membrane lead to forma-

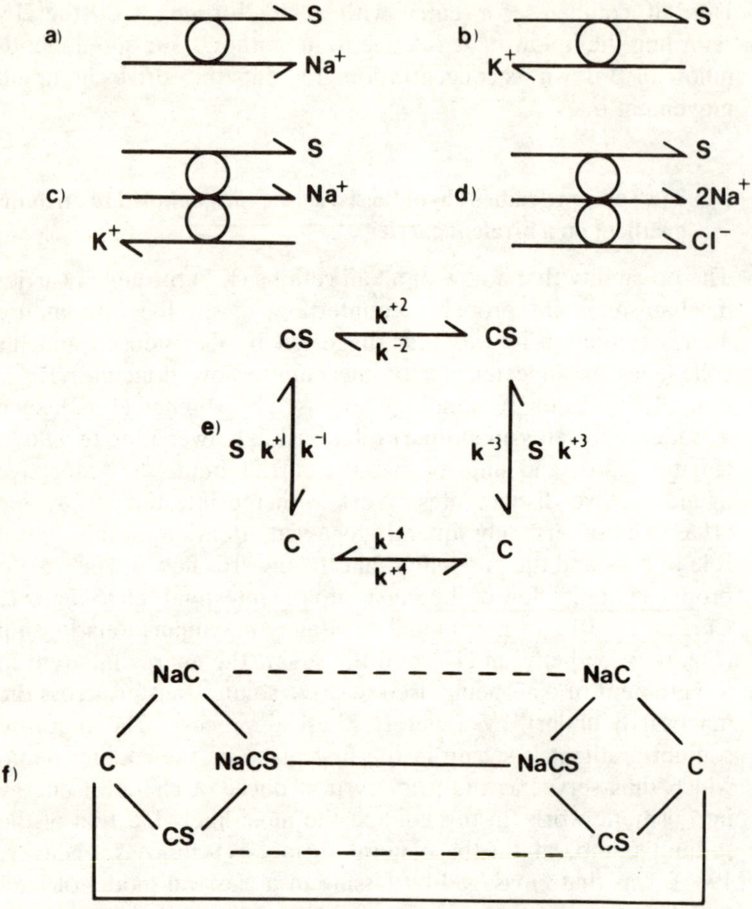

Fig. 2.2 Carrier models of the general relations between the flow of solute S and the flow of co-substrate ions. (*a*) symport with Na^+; (*b*) antiport with K^+; (*c*) electroneutral symport/antiport model; (*d*) electrogenic symport with $2Na^+$ and Cl^-; (*e*) simple carrier model; (*f*) bivalent carrier showing translocation of ternary complex NaCS and possibility of translocating the binary complexes NaC or CS. The large number of rate constants involved in the mathematical description of such carrier models leads to complex equations. One simplifying assumption is that binding or release of Na^+ or S is so much faster than other steps in the carrier cycle, the translocations, that binding is in quasi equilibrium. This assumption would be implausible if the ligands bind at sites buried in the membrane and both binding and translocation involve relatively slow changes in protein conformation.

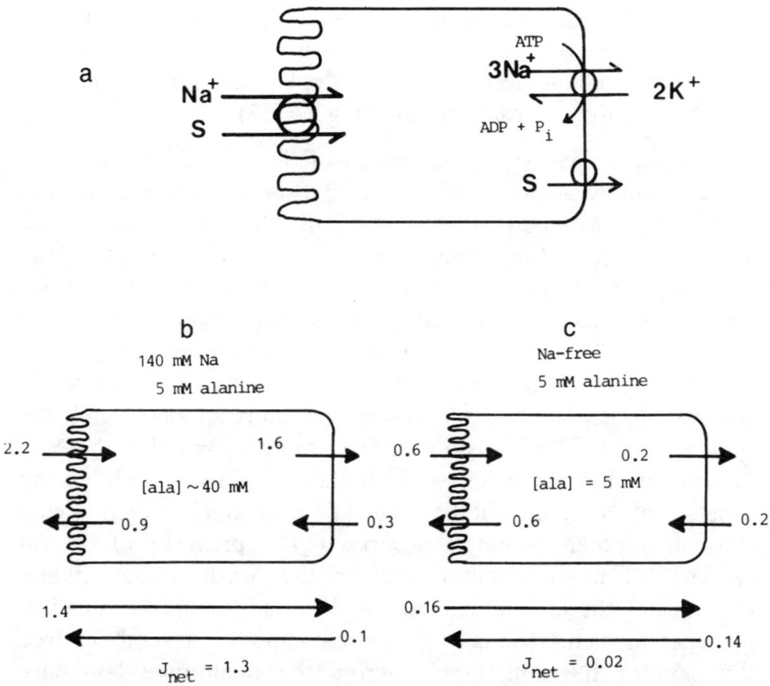

Fig. 2.3(*a*) Symport of Na^+ at the mucosal boundary during transfer of solute, S, across an epithelial cell. The Na^+ ions are subsequently expelled through the sodium pump at the baso-lateral face.

(*b*) The effect of removing Na^+ ions from the Ringer's solution bathing the mucosal and serosal faces of the rabbit ileum, on the undirectional fluxes (J) of alanine across these faces. J is in units of $\mu mol \cdot h^{-1} \cdot cm^{-2}$. (Adapted from Schultz, S. G. *et al.* (1967). *Journal of G General Physiology*, **50**, 1241–60 by copyright permission of the Rockefeller University Press.)

tion of a parallel concentration gradient of the substrate without the intervention of ATP?

(iii) Do inhibitors of the sodium pump also inhibit the accumulation and transfer of organic solutes?

2.3.1 *The requirement for Na^+*

An early study of the processes of L-alanine uptake at the respective mucosal and serosal faces of the rabbit small intestine showed that:

(i) the initial rate of amino acid uptake at the mucosal boundary;
(ii) the net transfer of alanine across the tissue; and
(iii) its intracellular accumulation all depended on the presence of Na^+ in the mucosal solution (Figure 2.3).

In contrast the passage of alanine across the serosal boundary was less affected by changes in the concentration of Na^+ at the mucosal surface (see Mircheff *et al.*, 1980 and Chapter 5). These observations received important confirmation from the study of the Na^+ dependence of amino acid uptake into membrane vesicles derived, on the one hand, from the brush border and, on the other hand, from the basolateral surfaces of the enterocyte (Sigrist-Nelson *et al.*, 1975; Lever, 1980). While Na^+ ions evidently stimulated alanine transfer across the mucosal boundary, the requirement for Na^+ was not in fact absolute. Whether the Na^+ dependent alanine carrier catalyses the transfer of alanine in the absence of Na^+, or whether the inflow of alanine then occurs through another system is uncertain. The presence of several carriers for neutral amino acids in the brush border greatly complicates the kinetic analysis, as does the seemingly universal presence in such systems of a kinetic component resembling free diffusion of the amino acid across the membrane boundary (Stevens *et al.*, 1984).

2.3.2 *The problem of co-substrate stoichiometry*

2.3.2.1 *A bivalent carrier.* A plausible model of the symport relationship between the inflow of Na^+ and that of the amino acid across the mucosal brush border envisages a bivalent carrier (C) associating with both ligands to produce the complex NaCS (Figure 2.2*f*). Let only the ternary complex NaCS translocate, in addition to C itself, so that equivalent amounts of Na^+ and S traverse the membrane together. The first attempts to test this possibility by Curran *et al.* (1967) showed that the ratio of the extra Na^+ absorbed to the equivalents of alanine simultaneously absorbed by the rabbit ileum was near 1.0 in the presence of 0.16 M Na^+, but that the ratio decreased systematically with the concentration of Na^+ in the mucosal solution. Alanine transfer appeared to occur both with Na^+, as mediated by the carrier complex NaCS, and through the same system without Na^+, as mediated by CS. However, a recent re-examination of the problem by Paterson *et al.* (1980) suggested that the transfer of alanine in the absence of Na^+ occurred through a second carrier system with which Na^+ did not interact, the amounts of alanine and Na^+ absorbed through the Na^+ dependent route being equivalent to each other, irrespective of the prevailing concentration of Na^+. The absorption of l-methionine, however, followed the more complex pattern whereby the Na^+-dependent carrier catalysed the transfer of methionine even in the absence of Na^+ (Paterson *et al.*, 1980). These observations serve to emphasise the complexity of the factors governing the stoichiometry of the respective flows of Na^+ and amino acids at the mucosal boundary. In contrast Na^+ was an oligatory co-substrate of 3-O-methyl-glucose absorption at the mucosal boundary of rabbit ileum, one equivalent of the substrate and co-substrate being absorbed together over a range of concentrations of each (Schultz, 1977).

2.3.2.2 *Tervalent carrier.* Vidaver (1964), in a penetrating study of the inflow of Na^+ ions accompanying the uptake of glycine by pigeon erythrocytes, concluded that the relevant carrier mechanism involved a quarternary complex 2NaCS, two Na^+ accompanying each amino acid equivalent into the erythrocyte.

The above considerations show that care is needed in interpreting the stoichiometry of the respective flows of substrate and co-substrate in terms of carrier models. The simplest case involves two special conditions:

(i) a single complex containing both ligands traverses the cell membrane whereas complexes containing only one type of ligand fail to do so;
(ii) ligand binding and release at either the cis or trans face is relatively fast in comparison with the translocation of the carrier complexes (Wheeler & Christensen, 1967; Kimmich & Randles, 1984).

A further important consideration in the assessment of carrier stoichiometry involves the changes in membrane potential that accompany solute absorption (Eddy, 1977; Kimmich & Randles, 1984).

2.3.3 *The initial rate of solute uptake as a function of [Na$^+$] or [S] and the 'degree' of the kinetic equations; the order and stoichiometry of ligand binding*

Given a system in which, on the one hand, the presence of Na^+ stimulates uptake of the substrate S and, on the other hand, the presence of S stimulates the uptake of Na^+, one can ask how the fluxes of each ligand depend on the respective concentrations of [Na^+] and [S]. The steady state solution of the equations describing the rate of uptake of S (V^S) as a function of [Na^+] can be represented as the rational function (Wong, 1975)

$$V^S = \frac{(\alpha_0 + \alpha_1[Na^+] + \alpha_2[Na^+]^2 + \alpha_3[Na^+]^3 \ldots)}{(\beta_0 + \beta_1[Na^+] + \beta_2[Na^+]^2 + \beta_3[Na^+]^3 \ldots)} \quad (2.2)$$

where the various coefficients α and β are constants at a given value of [S]. A similar expression, with $\alpha_0 = 0$ would apply to the flux of Na^+ itself. In eqn. (2.2) the degree of the polynomial terms with respect to [Na^+] is determined by the number of distinct ways in which Na^+ binds to the carrier (Wong, 1975). For instance, formation of NaCS by a random mechanism occurs by the addition of Na^+ either to C itself or to CS (Figure 2.2*f*). The maximum degree expected with respect to [Na^+] for the random binding model is thus two. Furthermore, because the formation of NaCS necessarily involves at least one step in which Na^+ is added, the minimum degree that can be expected is one. Indeed, for rather subtle reasons, the random binding model exhibits a degree of one when ligand binding on the cis and trans sides is rapid enough for equilibrium with the respective solutions to be reached. Equation (2.2) then takes the simpler form

$$V^S = (\alpha_0 + \alpha_1 [Na^+])/(\beta_0 + \beta_1 [Na^+]) \quad (2.3)$$

where the term α_0 represents the possibility that uptake of S does not fall to zero in the absence of Na^+ because CS translocates. When CS does not translocate α_0 is thus zero and V^S shows a simple hyperbolic dependence on [Na^+]. An equation of the same form applies when Na^+ and S bind to the carrier in that order either with or without the process reaching equilibrium.

Because there are several examples in the literature of Na^+

dependent amino acid transport which conform, at least approximately, to eqn. (2.3) with $\alpha_0 = 0$ the quasi-equilibrium assumption of ligand binding to the carrier is often assumed to apply. The scatter of the experimental observations in many such instances, however, probably precludes detection of minor second or third order terms in the rate equations. Another complicating factor, already noted in connection with the Na^+/alanine stoichiometry, is the possible presence of more than one carrier system. It is readily shown that two such systems acting together, each exhibiting a hyperbolic dependence on [Na^+], might conform algebraically to the version of eqn. (2.2) of second degree. The limitations of kinetic arguments based on the properties of molecular mixtures are indeed apparent but they are not easy to circumvent with the techniques at present available for fractionating membrane proteins.

The arguments set out above are concerned with the dependence of the rate of uptake of S on the concentration of the co-substrate ion [Na^+]. In a system where S and Na^+ only translocate together as NaCS, certain of the coefficients α and β of eqns. (2.2) and (2.3) are themselves functions of the substrate concentration [S]. Consider again the simple case where ligand binding occurs in random order and is in quasi equilibrium, translocation of C and NaCS being rate limiting and the trans concentration of Na^+ is negligible. In these circumstances the initial rate of uptake of S varies with [Na^+] and [S] as follows

$$V^{S} = (A[Na^+]\,[S])/([Na^+]\,[S] + B\,[Na^+] + C\,[S] + D) \quad (2.4)$$

where A, B, C and D are parameters determined solely by the magnitudes of the individual rate constants of the carrier cycle. Inspection of eqn. (2.4) reveals that the apparent K_m and V_{max} in such a system, with respect to either [S] or [Na^+], varies characteristically with the concentration of the other ligand (compare Figure 2.6). Thus the demonstration of such effects can be taken as indirect evidence for the formation of the ternary complex NaCS and, provided the uptake of both S and Na^+ are separately studied, as evidence of the translocation of the two partners together.

A first degree dependence of V^{S} on [Na^+] and [S] is also shown by models in which the ordered binding of the two ligands occurs rather than the random mode. Dependent on whether Na^+ or S binds first and whether binary carrier complexes translocate, K_m and V_{max} vary in a characteristic manner with the ligand concentration (Schultz & Curran, 1970). The formal extension of these approaches to systems where ligand binding is not at equilibrium and where binary carrier species also translocate obviously leads to considerable mathematical complexity.

Another important possibility is that more than one Na^+ translocates with the solute. Equation (2.2) is then at least second degree with respect to $[Na^+]$, a situation which occurs when the substrate and co-substrate are in binding equilibrium at both the cis and trans faces of the carrier.

2.3.4 *Inflow and outflow kinetics*

The properties referred to above involve the simplifying assumption that the substrate and co-substrate are both absent from the trans face of the membrane. A further test of the properties of real carrier systems accordingly consists in studying the flow of isotopically labelled substrate or co-substrate in relation to the trans concentrations of either species, especially when the system as a whole is in chemical equilibrium. Such tests have led to further distinctions between the random and ordered modes of binding of the substrate and co-substrate. Hopfer & Liedtke (1981) have attempted to compare on that basis the order in which Na^+ and glucose bind to the outer face of intestinal vesicles with the order in which they are released on the inner face (see also Semenza *et al.*, 1984). It is interesting in this connection that both the Na^+ dependent amino acid carrier of mouse ascites tumour cells (Johnstone, 1979) and the glucose carrier of the human erythrocyte translocate solute more rapidly in one direction than the other when the solute is present on only one side of the plasma membrane.

2.3.5 *Interactions with K^+ ions*

Studies by Crane and others, referred to on p. 68, indicated that extracellular K^+ antagonised the stimulatory role of Na^+ (Frizzell & Schultz, 1970). One possibility is that K^+ might compete with Na^+ for occupancy of the carrier (Figure 2.2*f*), thereby hindering the formation of the ternary complex NaCS. An extension of this idea is to suppose that K^+ serves as a co-substrate with an antiport role so that the symport role of Na^+ and the antiport role of K^+ occur in the same system (Figure 2.2*c*).

2.3.6 *Interactions with chloride ions*

Vidaver found that the Na^+ dependent uptake of glycine by pigeon erythrocytes was accelerated in the presence of chloride ions in a manner suggesting that these were co-factors and possibly co-substrates that were translocated in the mechanism. A number of other examples of this behaviour have been reported (see Nelson & Rudnick, 1982; Boge *et al.*, 1983; and Chapters 4 and 5). The likelihood is that glycine translocation in the above instance involves a carrier species liganded to two Na^+ ions, one Cl^- ion and one glycine molecule (Figure 2.2*d*).

2.3.7 *Interactions with the transmembrane electric field: conductance changes and membrane depolarisation associated with Na^+ dependent solute uptake*

Reference to Figure 2.2*a* shows that the symport of Na^+ with an uncharged solute S would tend to make the cellular phase electrically positive with respect to the extracellular phase representing the mucosal solution. In contrast, an antiport coupling to K^+ (Figure 2.2*b*) would tend to make the cellular phase relatively more negative, whereas an influx of co-substrate Na^+ that was obligatorily coupled to the efflux of K^+ (Figure 2.2*c*) would have no immediate effect on the membrane potential. Early work by Schultz and his colleagues showed that the active intestinal transport of sugars and amino acids increased the rate of transmural sodium transport across the short-circuited rabbit ileum by an amount roughly equivalent to the rate of solute transport (cf. Figure 2.4*a*). This behaviour is most readily explained if the Na^+ not merely activates the alanine carrier but serves as a co-substrate. Nevertheless, because the passage of Na^+ that is responsible for the short-circuit current occurs through the sodium pump (Figure 2.3*a*), the existence of this ouabain-sensitive, short-circuit current is not direct evidence that the Na^+/alanine symport itself transfers electric charge. Proof of this has come from other observations with intestinal or renal preparations

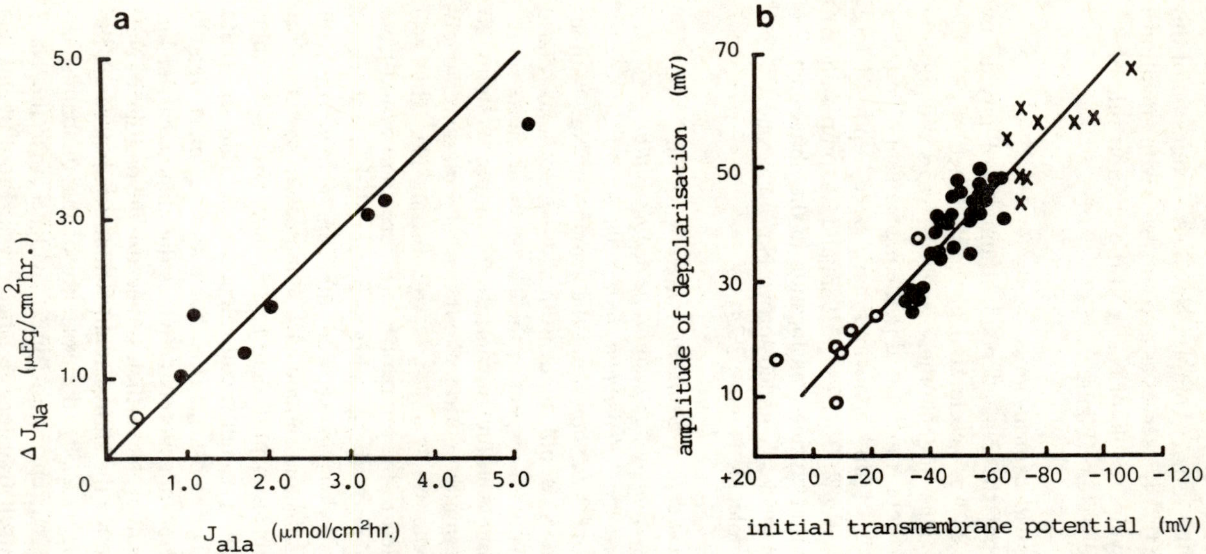

Fig. 2.4(*a*) The increase in sodium flux (J_{Na}) as a function of the net transepithelial alanine flux (J_{ala}) across short-circuited *Necturus* small intestine. (Taken with permission from Gunter-Smith, P. J. *et al.* (1982). *Journal of Membrane Biology*, **66**, 25–39.)

(*b*) Relationship between the initial intracellular potential and the amplitude of depolarisation caused by introducing alanine into the lumen of *Triturus* proximal tubules. The initial potential varied naturally or as a result of current injection. (Taken with permission from Hoshi, T. (1976). In: *Amino acid transport and uric acid transport*, eds. S. Silbernagl, F. Lang & R. Greger, pp. 96–104. Georg Thieme, Stuttgart.)

which demonstrated that Na^+ dependent solute influx depolarised the electrical potential across the apical membrane even under conditions where the epithelium was depleted of ATP and the activity of the sodium pump was impaired (Rose & Schultz, 1971; compare White & Armstrong, 1971). The amplitude of the depolarisation varied systematically with the magnitude of the membrane potential (Figure 2.4*b*). This behaviour would be expected if the inflow of substrate and co-substrate was electrogenic. The interpretation of these effects in terms of conductance changes is complicated by the well-established finding that Na^+ dependent solute transfer also leads to an increase in the transmural potential. Furthermore, the presence of low resistance pathways for electrical current between the epithelial cells means that electrical coupling occurs between the apical (mucosal) and basolateral (serosal) membranes. Analysis of the equivalent electrical circuits has led to the view that the inflow of Na^+ as co-substrate is indeed conductive, being associated with an immediate decrease in the effective resistance of the mucosal membrane. This change is followed by a decrease in the effective resistance of the serosal membrane as the sodium pump becomes activated (Gunter-Smith *et al.*, 1982)*. Sodium ion dependent current transfer during amino acid absorption has also been detected in pancreatic acinar cells and in oocytes. In the latter study the membrane depolarization was compensated by current injection and, under voltage-clamp conditions, the Na^+ dependent transfer of alanine was associated with the transfer of two equivalents of positive charge, behaviour corresponding probably to a 2:1 stoichiometry for the sodium-alanine co-transport mechanism. Co-substrate dependent membrane depolarisation has also been detected qualitatively by means of voltage-sensitive fluorescent dyes.

2.3.8 *Voltage-sensitive carrier models*

Much of the earlier work on the kinetics of solute transfer was done without taking into account the presence of an electric field

---

*These observations turn out to be intimately related to the changes in K^+ conductance of the basolateral membrane which are discussed with respect to the pancreas in Chapter 3.

in the cell membrane (Schultz & Curran, 1970). The transfer of Na^+ ions across the mucosal boundary, for instance, occurs between two solutions differing in electrical potential by about 40 mV ($\Delta\psi$ = −35 to −50 mV) (Freel & Goldner, 1981). The extracellular and intracellular interfaces are also characterized by their respective surface potentials, changes in which can lead to complex kinetic effects. The electric field might influence the different steps in the carrier cycle in at least three ways.

(i) The free energy barrier impeding carrier translocation in the membrane might be regarded as a single symmetrical peak (Britton, 1965). Let *k* be the rate constant in the absence of an electric field, corresponding to the transfer of the carrier complex NaCS, bearing a unit formal positive charge, from the cis to the trans location. Let the factor $e^{-\Delta\psi F/RT} = \theta^2$. Then the rate constant for the above translocation in presence of a membrane potential $\Delta\psi$ is θk, whereas that for the reverse translocation is multiplied by $1/\theta$. The membrane depolarisation associated with solute uptake might on that view lead to changes in certain translocation rate constants in the order of a factor of 2 or 3. The fact that many workers have observed simple hyperbolic dependence of solute uptake on substrate or co-substrate concentration (eqn. 2.4) without considering the membrane potential as an independent variable limits the utility of such observations as the basis for constructing kinetic models.

(ii) An alternative treatment of the kinetic effect of changes in $\Delta\psi$ is to consider that the Na^+ ion is selectively drawn into a 'well' under the influence of the electric field. In effect the energy inherent in that field is converted into an enlarged ionic gradient of Na^+ (Mitchell, 1977).

(iii) Another procedure* is to regard the electric field as causing conformational changes in the carrier that alter the magnitude of the rate constants defining the binding or release of

---

*While different models are all constrained by the requirement for their rate constants to conform to the principle of microscopic reversibility (see Britton, 1965), they differ in the way the electric field affects individual rate constants including those governing the binding of ligands that are not translocated (Semenza *et al.*, 1984).

the substrate or co-substrate from the carrier (Aronson, 1978).

These different ways of describing how the rate of solute transport is affected by changes in the membrane potential correspond to different physico-chemical models of the carrier mechanism, the molecular basis of which is largely unknown.

2.3.9 *Kinetic and thermodynamic description of coupling between the flows of substrate and co-substrate*

If the study of the interdependence of the initial rate of uptake of the substrate and the putative co-substrate, in the presence of various concentrations of both ligands, suggests that the two ligands move together across the cell membrane, the question next arises whether the coupling between the two flows is linked mechanistically, as the co-substrate hypothesis requires, or whether these flows simply vary in parallel for other incidental reasons. For instance, the properties of the respective channels through which Na^+ and S traverse the cell membrane might be independently affected by the presence of the other ligand, even though the two flows were not directly coupled. In that event the flow of Na^+ would be determined only by the magnitude of its own electrochemical gradient ($\Delta\mu^{Na}$) acting across the cell membrane, but not by the corresponding electrochemical gradient of S ($\Delta\mu^{S}$). In the formalism of irreversible thermodynamics

$$J^{S} = L_{S}\,\Delta\mu^{S} \text{ and } J^{Na} = L_{Na}\,\Delta\mu^{Na} \tag{2.5}$$

where the flows of both species are determined by the product of the conjugate force and the magnitude of the respective phenomenological coefficients (L_S and L_{Na}). In either case the spontaneous flow of material occurs down the electrochemical gradient of the relevant species. This situation contrasts with the equations describing the flows of substrate and co-substrate according to the simplest version of the sodium gradient hypothesis. The respective flows of S and Na^+ now depend both on $\Delta\mu^{Na}$ and $\Delta\mu^{S}$ as follows

$$J^{S} = L_{S}\,\Delta\mu^{S} + L\,\Delta\mu^{Na} \tag{2.6}$$

$$J^{Na} = L_{Na}\,\Delta\mu^{Na} + L\,\Delta\mu^{S} \tag{2.7}$$

where L is a cross-coefficient, conforming to the so-called Onsager reciprocal relation and representing the coupling between the movements of the two types of particles.* Two important predictions are thus as follows.

(i) The flow of S or of Na^+ is governed not merely by its own gradient but also by that of its partner. A 'downhill' gradient of the one ligand can therefore lead to an 'uphill' gradient of the other ligand. For instance, when $\Delta\mu^S = 0$ a gradient of Na^+ that causes the spontaneous inflow of that ion will simultaneously result in a spontaneous inflow of S.

(ii) The flow of S is governed solely by its own gradient and by that of the co-substrate, no other source of energy being required.

2.4 Solute accumulation as a function of $\Delta\mu^{Na}$ at thermodynamic equilibrium

(i) When the accumulation of S occurs solely at the expense of the energy inherent in the sodium gradient, a constant n moles of Na^+ being transferred per mole of S, thermodynamic equilibrium is reached when

$$n\,\Delta\mu^{Na} + \Delta\mu^{S} = 0 \tag{2.8}$$

The magnitude of the solute gradient that is in thermodynamic equilibrium with the co-substrate gradient is thus larger the bigger the value of n. Maintenance of a strict stoichiometry of n equivalents of Na^+ absorbed per mole of S absorbed requires that intermediate carrier species, NaC or CS, etc., do not translocate.†

---

*An illuminating discussion of the kinetic basis of the phenomenological coefficients occurs in Hill (1982).

†Naftalin (1984) has recently argued that eqn (2.8) leads to paradoxical results if two symport mechanisms differing in stoichiometry are imagined to function in series. The first ($n = 2$) would equilibrate the substrate and co-substrate between solution 1 and solution 2 and the second ($n = 1$)

(ii) A similar argument applies to carrier models in which translocation of the intermediate species, NaC, etc., is not forbidden. Because these involve closed cycles of reactions in which only Na^+ or S translocate, equilibrium is only achieved when the substrate gradient and the co-substrate gradient are each separately zero. However, as outlined below, the solute accumulation ratio can be maintained at other steady values in the presence of a finite electrochemical gradient.

2.4.1 *The steady state of solute accumulation maintained by dissipation of the sodium electrochemical gradient: slippage*

When $J^S = 0$ in eqn (2.6)

$$\Delta\mu^S = -(L/L_S)\,\Delta\mu^{Na} \tag{2.9}$$

However, because J^{Na} is not then zero, the maintenance of a steady gradient of S requires the continuing passage of Na^+ down its own gradient. As the driving gradients ($\Delta\mu^S$, $\Delta\mu^{Na}$) vary, the stoichiometrical ratio of J^S and J^{Na} (flow ratio) in such models is not constant even though the coefficients L_{Na}, L_S and L remain constant. One interesting situation is where $\Delta\mu^{Na}$ is constant and $\Delta\mu^S$ is varied with the result that J^{Na} and J^S also vary. It can then be shown (see, for example, Heinz, 1978) that

$$(\delta J^{Na}/\delta J^S)_{\Delta\mu^{Na}} = -(\Delta\mu^S/\Delta\mu^{Na})_{J^S=0} \tag{2.10}$$

The importance of this approach is that it shows, within the limitations of the validity of the phenomenological equations, that

---

would equilibrate their distributions between solutions 2 and 3. The paradox arises because the ligand concentrations can be chosen in such a way that solutions 1 and 3 are evidently not in equilibrium with each other. The origin of the paradox, however, appears to be not that eqn 2.8 is wrong but that a notional transfer of S and Na^+ between solutions 1 and 3, *via* 2, leads to accumulation of Na^+ in solution 2. There can be no question therefore of solutions 1 and 3 being equilibrated by the operation of such a device. Hill's carrier model advocated by Naftalin (1984) envisages the possibility of binary translocations and therefore obeys eqn (2.9).

the differential flow ratio, as defined, and the maximum accumulation ratio ($[S]_i/[S]_0$) are closely related quantities. The parallel between eqns (2.10) and (2.8) is noteworthy. A comparison of the kinetic and thermodynamic approaches to computing the way the solute accumulation ratio varies with 'slippage' is illustrated in Table 2.2. (The general condition that 'slippage' occurs is that the substrate and the co-substrate can traverse the carrier through at least two routes differing in stoichiometry, e.g., $2\,Na^+/S$ and $1\,Na^+/S$ or $0\,Na^+/S$.) Neither the solute flow ratio at any one value of $\Delta\mu^{Na}$, nor the solute accumulation ratio it maintains provide a direct measure of the stoichiometry of the carrier complex in the absence of other kinetic information (p. 56).

2.4.1.1 *Passage of electric charge through the symport* As noted above, several studies have shown that amino acids or carbohydrates cause rapid membrane depolarisation, the inward current probably being carried by Na^+ ions (Petersen, 1982 – see Petersen, 1986); Petersen showed that the apparent reversal potential (compare Figure 2.4*b*), at which the amino acid would cause no decrease in the initial membrane potential was in the region of +30 mV, in a system where the Nernst potential for Na^+ itself was about +75mV. The significance of the amino acid reversal potential defined in that way is obscure, because the relevant magnitude of the amino acid gradient is not known. The ratio $\frac{RT}{F}\ln\frac{[Na^+]_0[S]_0}{[Na^+]_i[S]_i}$ would initially correspond to an infinitely large positive reversal potential, if the flow of co-substrate ions alone governed the membrane potential.* An interesting avenue for further investigation would be to carry out such experiments with preparations in which the current through the system at a given voltage was studied as a function of both the amino acid gradient and the sodium concentration gradient. It might be possible in that

*The theoretical point being made here should be compared with the observations on the pancreas. The question of the reversal potential for the amino acids is not straightforward since the amino acid concentration inside and outside the cell must be important in addition to the equilibrium potential for the relevant ion which is co-transported (Chapter 3.4).

Table 2.2 *Computation illustrating the effect of increasing the mobility of the binary complex NaC relative to that of NaCS on the flow ratio* (J^{Na}/J^{S}) *and the maximum accumulation ratio* ($[S]_i/[S]_0$)

Slippage factor k'/k	Flow ratio	$[S]_i/[S]_0$	$(27)^x$
0	1.00	27.0	27
0.01	0.97	22.2	25
0.05	0.88	13.2	18
0.20	0.62	5.7	7.7
0.50	0.31	3.1	2.8

The procedures used are described in Eddy (1968*a*) and Hacking & Eddy (1981). The carrier C was assumed to bear a single negative formal charge. The mobility of C was computed as $k\theta$ in one direction and k/θ in the other (p. 62). The ratio of the mobility of NaC to that of NaCS is described by the factor k'/k below. The increase in the current of Na^+ through the carrier in the presence of S (J^{Na}) was computed as was the flow (J^S) of S itself. $[Na]_0$ was 150 mM, $[Na]_i$ was 50 mM, $[S]_0$ was 60 mM, Θ^2 was 9. A single dissociation constant of 20 mM was assumed to characterise both the association of Na^+ with C and the subsequent association of NaC with S. Comparison with the accumulation ratio expected on the basis of eqn 2.10 was made by computing the value of $(27)^x$ where x = flow ratio.

way to assay the magnitude of any 'slippage' currents associated with the Na^+ dependent amino acid carrier.

2.4.2 *Effects of potassium ions or chloride ions on solute accumulation at the expense of the sodium electrochemical gradient*

When uptake of an uncharged solute is accompanied by the uptake of one equivalent of Na^+ compulsorily linked to an outflow of one equivalent of K^+, the position of thermodynamic equilibrium is independent of $\Delta\psi$ and occurs when

$$\mu_0^{Na} + \mu_0^{S} + \mu_i^{K} = \mu_i^{Na} + \mu_i^{S} + \mu_0^{K} \qquad (2.11)$$

Symport with Cl^- can be described in a similar way. An interesting early view of the mechanism of solute uptake linked to the flow of Na^+ was that the energy inherent in that part of the sodium gradient represented by the electric field, as opposed to the concentration gradient of Na^+, might not be coupled to the transport process but simply dissipated (Schultz & Curran, 1970).

However, coupling to the flow of Na^+ alone, unless S itself bears an equal and opposite charge, of necessity leads to coupling to the electric field across the cell membrane. This is because, in the presence of an electric field, association of the carrier with Na^+ affects the distribution of the various carrier species between their inwards and outwards configurations. It is this distribution which governs the maximum solute gradient that the sodium gradient maintains.

A related matter concerns the possible role of K^+ ions in driving solute accumulation. Crane (1977) and Schultz & Curran (1970) argued that if K^+ inhibited the function of Na^+ in accelerating solute uptake, the normal distribution of K^+ across the cell membrane would tend to inhibit Na^+ dependent efflux of a solute more than the corresponding solute influx. It was held that the net effect would be that the gradient of K^+ would increase solute accumulation, even though K^+ ions themselves were not traversing the cell membrane (see Crane, 1977). However, solution of the carrier equations for a system interacting with Na^+, K^+ and S shows that this effect is illusory. In the absence of an actual coupled movement of K^+ across the cell membrane through the carrier, the distribution of K^+ has no effect on the steady state distribution of the solute that the gradient of Na^+ maintains. The thermodynamic principle involved is that utilisation of the energy inherent in $\Delta\mu^K$ requires that K^+ ions flow down that gradient (Eddy, 1978). These considerations do not rule out the possibility that K^+ ions, without acting as a co-substrate, might in some other way change the extent to which the symport mechanism coupled the flows of Na^+ and S across the cell membrane (see p. 80).

2.5 Demonstration of solute accumulation driven by the sodium gradient without the intervention of metabolic energy

Experimental tests of the sodium gradient hypothesis centre about the following questions (Schultz, in Andreoli *et al.* 1978; Kimmich, 1981).

(i) Do cellular preparations or membrane vesicle preparations depleted of ATP in fact concentrate specific solutes when $\Delta\mu^{Na}$ is appropriately manipulated (see Figure 2.5) and is the

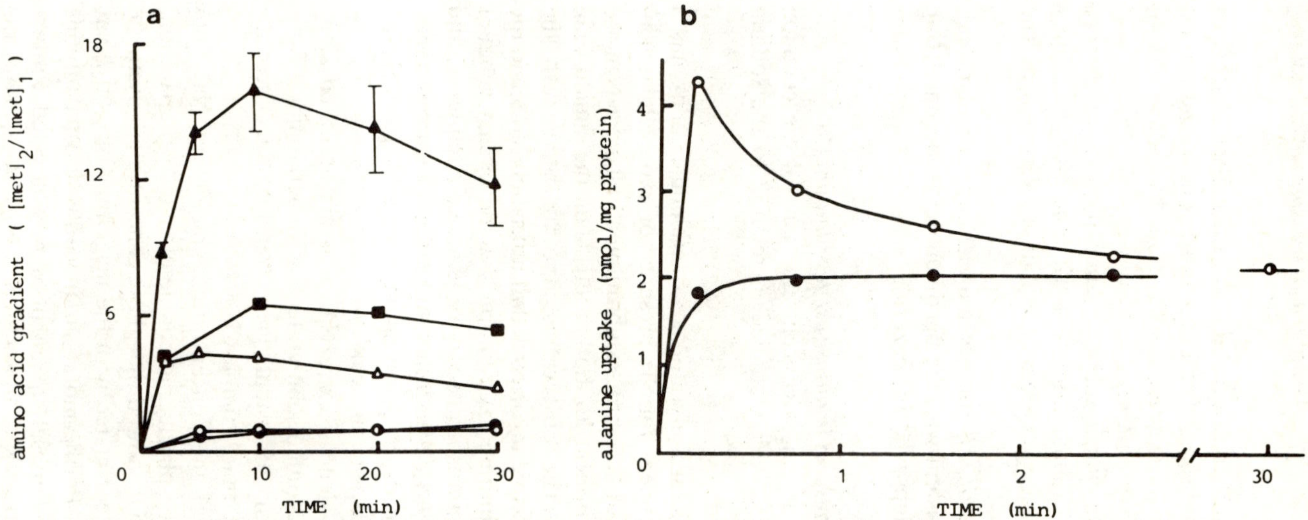

Fig. 2.5 Effect of increasing $\Delta\mu^{Na}$ on amino acid accumulation. (*a*) Mouse ascites tumour cells depleted of ATP accumulated methionine in the presence of valinomycin and a graded membrane potential governed by the extracellular concentration of K^+. The latter was 5 mM (▲), 22 mM (■) or 115 mM (●). When [K^+] was 5 mM the presence of valinomycin (▲) stimulated amino acid uptake several fold as compared with a control lacking it (Δ). No stimulation occurred when [K^+] was 115 mM (0, ●). (Reprinted by permission from Reid, M. *et al.* (1974). *Biochemical Journal*, **140**, 383–93. Copyright © 1974 The Biochemical Society, London.) (*b*) Uptake of L-alanine by brush border membrane vesicles from the rat intestine was assayed either in the presence of a gradient of NaSCN (0) or when that gradient had already dissipated (●). (Taken with permission from Sigrist-Nelson, K. *et al.* (1975). *Journal of Biological Chemistry*, **250**, 5674–80.)

sodium gradient then dissipated faster than in the absence of the solute? (As discussed in Chapter 4.1, one of the earliest examinations of these questions was made by Vidaver (1964) using pigeon erythrocytes.)

(ii) Is the extent to which solute accumulation occurs, under the former conditions, consistent with the way that solute accumulation during energy metabolism varies with the magnitude of $\Delta\mu^{Na}$?

(iii) Is the flow ratio (J^{Na}/J^{S}) observed under any particular set of conditions consistent with the maximum solute accumulation ratio that develops when J^{S} is zero?

2.5.1 *Amino acid transport in mouse ascites-tumour cells*

The ability of this system to concentrate many of the common amino acids has been widely studied in relation to the sodium gradient hypothesis.

(i) Glycine, 2-aminoisobutyrate or L-methionine influx has a marked requirement for the presence of extracellular Na^+ ions (Figures 2.5 and 2.6), exhibiting first degree saturation kinetics with respect to both $[Na]_0$ and the amino acid concentration. The observations are consistent with the presence of a bivalent carrier to which one Na^+ ion and the amino acid both bind to equilibrium as the ternary complex NaCS. This is the so-called A system (see Chapter 1, and Christensen, 1984). The influx and efflux of various amino acids also proceeds, relatively slowly, through a parallel Na^+ independent pathway or 'leak'.

(ii) About 1 equivalent of extra Na^+ ions are absorbed along with glycine, 2-aminoisobutyrate and L-methionine (Table 2.3), behaviour consistent with translocation occurring mainly as the ternary complex NaCS.

(iii) Manipulation of the concentration gradient of Na^+ ions acting across the plasma membrane of cellular preparations depleted of ATP showed that these concentrated glycine to an extent that correlated with the magnitude of $[Na]_0/[Na]_i$ (Figure 2.7*b*). Furthermore, an inwardly directed glycine gradient tended to set up an outwardly directed concentration gradient of Na^+ ions (Eddy, 1969). Analysis of the

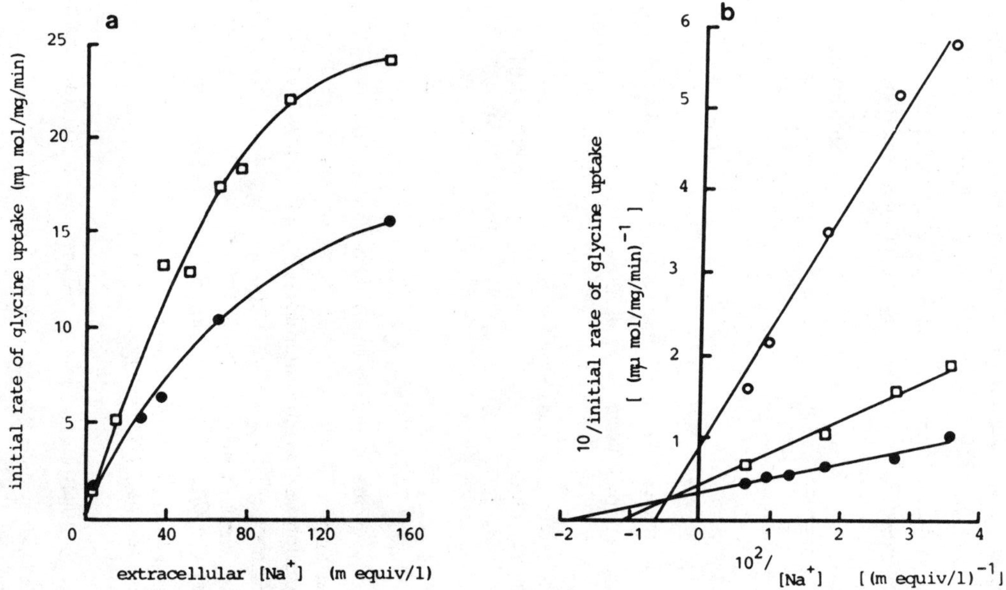

Fig. 2.6 Rate of glycine uptake by mouse ascites tumour cells as a function of the extracellular concentrations of Na^+ ions and of glycine itself. (*a*) Uptake at 37°C from Ringer solutions containing 4 mM glycine (●) or 12 mM glycine (□). (*b*) Reciprocal of the initial rate of glycine uptake depicted as a function of the reciprocal of extracellular [Na^+] at glycine concentrations of 1 mM (○), 4 mM (□) or 12 mM (●). (Reprinted by permission from Eddy, A. A. *et al*. (1967). *Biochemical Journal*, **103**, 863–76. Copyright © 1967 The Biochemical Society, London.)

Table 2.3 *Observations bearing on the stoichiometry of the coupling between the flow of Na^+ ions and organic solute through the symport mechanism*

System	Stoichiometry
Ratio of solute dependent Na^+ flux to Na^+ dependent solute flux (flow ratio)[a]	
Glycine in pigeon erythrocytes (Vidavor, 1964)	1.5–2.2
Alanine in hepatocytes[b] (Kristensen & Folke, 1983)	1.0
Alanine in rabbit ileum (see text and Paterson *et al.*, 1980)	0.94±0.04
Alanine in voltage-clamped *Necturus* small intestine (Figure 2.5)	1.0
Aminoisobutyrate in mouse ascites tumour cells	
(Schafer & Jacquez, 1967)	0.85–1.05
(Dawson *et al.*, 1983)	0.90±0.02
(Yamamoto & Kawasaki, 1981)	0.8–1.0
Glycine in mouse ascites tumour cells (Eddy, 1968*a*; Philo & Eddy, 1978*b*)	0.90±0.11
Methionine in mouse ascites tumour cells[b] (Eddy, 1985)	1.0–1.3
3-O-methylglucose uptake in pigeon enterocytes[b] (Kimmich & Randles, 1984)	1.7–2.3
3-O-methylglucose in rabbit ileum (Schultz & Curran, 1970)	0.9–1.2
D-glucose in voltage-clamped pig renal cells (Misfeldt & Sanders, 1982)	2.1
D-glucose in bovine intestinal vesicles[b] (Kaunitz & Wright, 1984)	2.0–2.3
D-glucose in rabbit kidney outer cortical vesicles[b] (Fukuhara & Turner, 1984)	1.0
Ratio of current change to associated flow of solute	
Glucose-induced short circuit current in pig renal cells (Misfeldt & Sanders, 1982)	2.0
Current induced by alanine in voltage-clamped oocytes (Jung *et al.*, 1984)	1.6–2.1

Table 2.3 (*continued*)

Ratio $\Delta\mu^S/\Delta\mu^{Na}$ when $J^S = 0$ (see also Fig. 2.10) System	Stoichiometry
Rabbit kidney outer medullary vesicles with glucose (Fukuhara & Turner, 1984)	1.5–1.8
Rabbit kidney outer cortical vesicles with glucose (Fukuhara & Turner, 1984)	0.9–1.0

[a]Relationship to composition of carrier complex $n\text{Na}^+\text{CS}$ indirect unless 'slippage' is assumed to be small (see text). Corrections required for membrane depolarisation unless voltage-clamped preparations are used (Eddy, 1977; Kimmich & Randles, 1984). Other corrections for substrate cycling are discussed in Philo & Eddy (1978*b*). A net uptake of Na^+ with the solute was explicitly demonstrated in several of the instances cited.
[b]Voltage clamped in presence of valinomycin

observations illustrated in Figure 2.7*b* showed that the glycine gradient $[S]_i/[S]_0$ increased with $[Na]_0/[Na]_i$ by a larger factor than could be attributed to the increase in the sodium concentration gradient alone. The additional factor initially appeared to be the magnitude of the gradient of K^+, rather than a varying membrane potential, for the latter was believed to be consistently small (Aull, 1967). However, later work by Schafer & Heinz (1971) ruled out a simple Na^+ symport/K^+ antiport model (Figure 2.2*c*). Gibb & Eddy (1972) accordingly suggested that amino acid transfer was coupled to the electrochemical gradient of Na^+ ions, thereby focussing attention on the problematical question of the magnitude of the membrane potential ($\Delta\psi$) the tumour cells maintained in various circumstances.

(iv) An important advance was the recognition that the early estimates of $\Delta\psi$ using capillary microelectrodes were subject to a technical error whereby imperfect sealing of the implanted electrode tended to short circuit the membrane. When allowance is made for that factor potentials of about −27 mV are observed. Assays with a voltage-sensitive

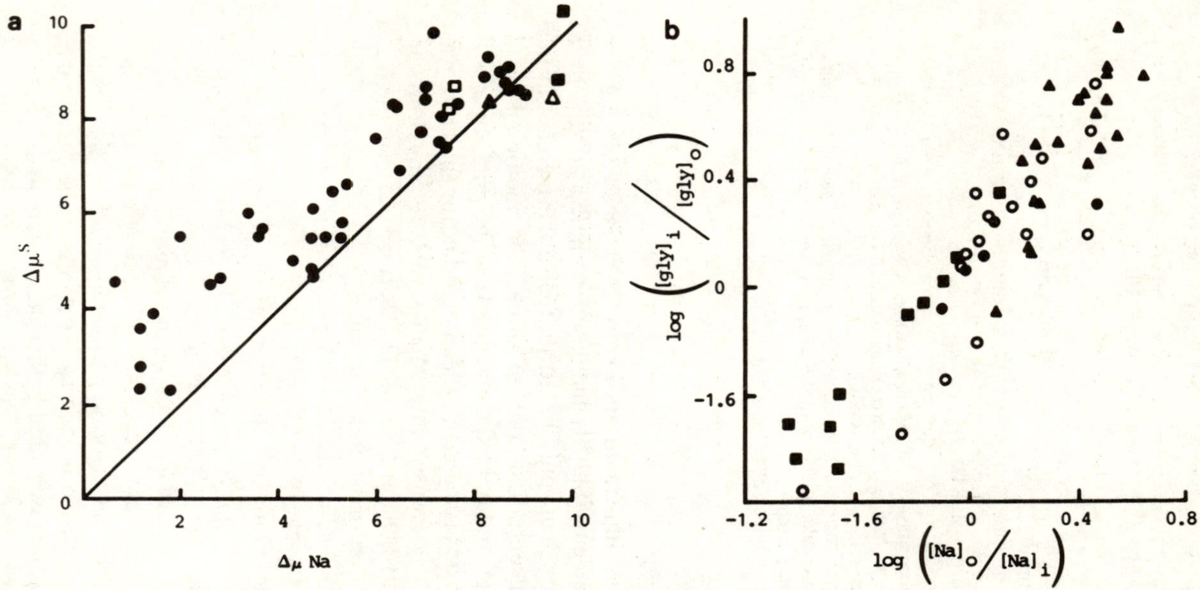

Fig. 2.7 Relative magnitudes of $\Delta\mu^S$ and $\Delta\mu^{Na}$. (*a*) The assays made in three laboratories are combined so as to compare the magnitude of the amino acid gradient which mouse ascites-tumour cells maintain during energy metabolism, in various conditions, with the corresponding magnitude of the sodium electrochemical gradient. (●) 1 mM methionine (Philo & Eddy, 1978b); (■), 0.1 mM or 1 mM 2-aminoisobutyrate (Hacking & Eddy, 1981); (□), 0.5 mM 2-aminoisobutyrate (Henius & Laris, 1979). (Δ), 1 mM glycine (Hacking & Eddy, 1981); (▲), 0.01 mM glycine (Hoffman & Lambert, 1983). Gradients are expressed in units of $kJ \cdot mol^{-1}$.

(*b*) This figure shows how the transient amino acid gradient formed by tumour cell preparations depleted of ATP varies with the sodium concentration gradient. (Reprinted by permission from Eddy, A. A. (1968*b*). *Biochemical Journal*, **108**, 489–98. Copyright © 1968 The Biochemical Society, London.)

carbocyanine dye gave values ranging from about $-36\,mV$ in the above cell line (Hoffmann & Lambert, 1983) to $-50\,mV$ in other tumour cells (Philo & Eddy, 1978a and b).*

By manipulating the magnitude of $\Delta\psi$ with valinomycin (Figure 2.5) and other ionophores, it proved feasible to reproduce, in cellular preparations in which energy metabolism had stopped, amino acid gradients of similar magnitude (8 kJ/mol) to those formed during energy metabolism (Reid *et al.*, 1974; Johnstone, 1979). Relatively small amino acid gradients also formed in vesicles in the presence of a favourable sodium gradient (Johnstone, 1979).

(v) Membrane depolarisation, indicative of a net cation influx associated with amino acid absorption, was detected by means of a voltage-sensitive dye (Johnstone, 1979) and is illustrated in Figure 2.8.

(vi) The way in which the amino acid gradient, formed at relatively low extracellular concentrations of various amino acids, varied with the magnitude of $\Delta\mu^{Na}$ is illustrated in Figure 2.7*a*. It is very striking that several of these preparations maintained a somewhat larger amino acid gradient than the corresponding sodium gradient. After correction for leak pathways, Philo & Eddy (1978b) and Hacking & Eddy (1981) estimated that, bearing in mind the flow ratio of 1 for substrate and co-substrate, the discrepancy corresponds to a factor of about 2 ($1.8\,kJ\,mol^{-1}$) over the whole range in which $\Delta\mu^{Na}$ varied. This factor may represent the effective activity coefficient for Na^+ ions in the cytosol (Freel & Goldner, 1981). If so, the sodium gradient appears to be energetically competent to maintain amino acid accumulation during energy metabolism.

(vii) The above observations have been taken to mean that, in accordance with eqn (2.8) (p. 64), the amino acid gradient and the sodium electrochemical gradient are near thermodynamic equilibrium in the steady state of amino acid

*Even though the use of fluorescent dyes, in different laboratories, to assay $\Delta\psi$ has led to estimates of $\Delta\mu^{Na}$ that are consistent (see Petersen, 1982), the procedures employed have been questioned (Smith, 1982) in the writer's view somewhat implausibly.

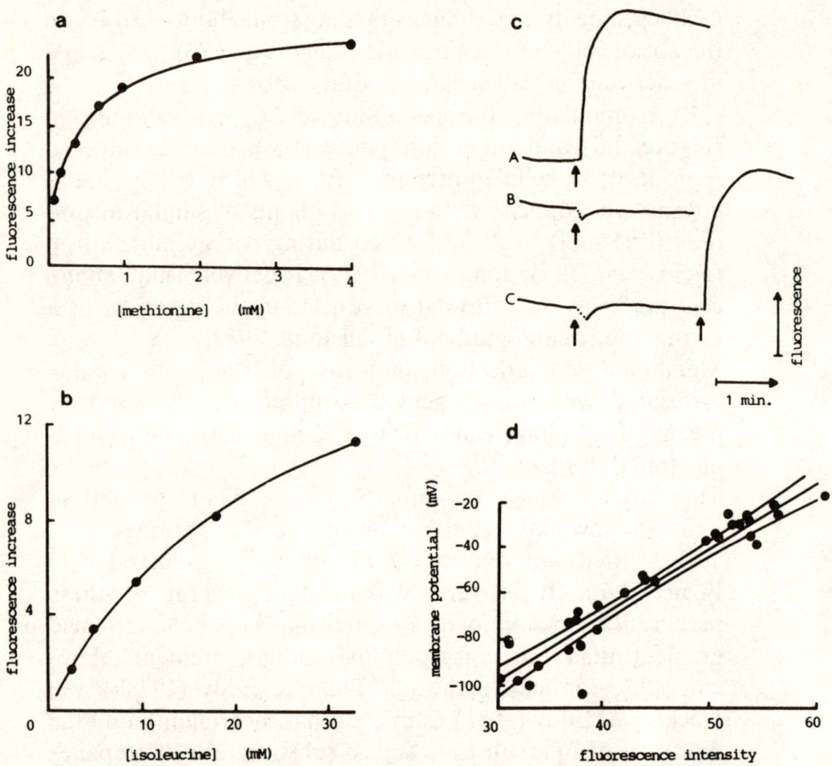

Fig. 2.8 The electrogenic character of Na^+-dependent amino acid absorption by mouse ascites-tumour cells, assayed with a voltage-sensitive carbocyanine dye (Philo & Eddy, 1978a; Hacking & Eddy, 1981). Increase in fluorescence (membrane depolarisation) caused by the addition of graded concentrations of (*a*) L-methionine or (*b*) L-isoleucine. The fluorescence increase only occurred in the presence of Na^+ ions: (*c*) shows that glycine depolarised the cells in the presence (A) of Na^+ ions but not in their absence (B). In contrast, the L System substrate, a norbornane amino acid had no significant effect on the fluorescence in the presence of Na^+ ions (C), the subsequent addition of glycine, however, causing depolarisation. In (*d*) the system is calibrated, the fluorescence intensity being shown as a function of the membrane potential, under the conditions in which assays such as those illustrated in Figure 2.7 were performed.

distribution that the tumour cells maintain in the presence of relatively dilute solutions (< 1 mM) of L-methionine, glycine or 2-aminoisobutyrate. However, it has long been known that provision of larger concentrations of amino acid results in the formation of much smaller amino acid gradients (Heinz & Patlak, 1960). This is apparently because a leak pathway bypasses the action of the amino acid pump rather than because the sodium gradient itself is greatly lowered (Hacking & Eddy, 1981). Indeed certain amino acids such as L-leucine and L-isoleucine, which exhibit a relatively low affinity for the A system of amino acid transport, fail to reach thermodynamic equilibrium with the sodium gradient even at the lower substrate concentration. Hoffmann & Lambert (1983) argue that the permeability of the tumour cell membrane to amino acids through the 'leak' pathway varies with cellular volume and is a component of the system controlling that volume.

2.5.2 *Is the sodium gradient the necessary and sufficient source of energy driving amino acid accumulation*?*

One objection to the arguments outlined above is that certain amino acids become concentrated in the tumour cells in the absence of an appropriate sodium gradient. For instance, Christensen *et al.* (1973) and others have shown that specific substrates of the transport system L are concentrated in the absence of Na^+. Certain norbornane amino acids which are not substrates of the A system behave in that way, as does L-isoleucine which is a good substrate of system L and a poor substrate of system A (Hacking & Eddy, 1981). The norbornane amino acid fails to depolarise the cell membrane (Figure 2.8*c*), is concentrated extensively by the tumour cells even when ouabain has dissipated the sodium gradient, and appears not to utilise an electrogenic co-substrate mechanism (Hacking & Eddy, 1981; Valdeolmillos *et al.*, 1982). It

---

*The problematical ability of certain redox reagents to accelerate amino acid uptake through system A appears to be due to an increase in the K^+ ion conductance of the cell membrane, rather than to direct energisation of amino acid absorption by redox reactions (Ohsawa *et al.*, 1980; Valdeolmillos *et al.*, 1982).

is uncertain whether the L system concentrates its substrates simply by exchanging them with cellular amino acids (Johnstone, 1979) or whether some other mechanism is involved (Ohsawa *et al.*, 1980).

Study of the absorption of 2-aminoisobutyrate in circumstances where the normal direction of the concentration gradient of Na^+ ions is reversed have been especially significant in relation to the question whether the sodium gradient alone energises the system. When ATP is lacking the amino acid then tends to be expelled from the cells, whereas during glycolysis these concentrate the amino acid (Schafer & Heinz, 1971). Gibb & Eddy (1972) attributed this latter behaviour to the circumstance that during glycolysis the initial flow of amino acid takes place under conditions where the sodium pump hyperpolarises the cell membrane sufficiently to create an inwardly directed sodium electrochemical gradient. Heinz *et al.* (1981) devised a stringent test of the gradient hypothesis by including nigericin in the above system, an ionophore that catalyses the electroneutral exchange of Na^+, K^+ and protons. The sodium concentration gradient then remains inverted throughout the experimental period, provided lactate rather than glucose is used as energy substrate (Figure 2.9), a membrane potential of the order of −90mV being required to explain the accumulation of amino acid in terms of the sodium gradient hypothesis. Application of a voltage-sensitive carbocyanine dye suggested that the membrane potential was quite insufficient being only about −30 mV (Heinz *et al.*, 1981). However, the use of this technique would inhibit oxidative phosphorylation thereby depolarising the tumour cells, and it seems more significant that the behaviour of the system in the presence of an oxonol probe gave results qualitatively consistent with the gradient hypothesis (Eddy, 1985).

2.5.3 *Related aspects of carbohydrate transport in renal or intestinal preparations*

The concentration of various solutes in renal or intestinal vesicle preparations, at the expense of the sodium electrochemical gradient, is a well-established phenomenon (Lever, 1980; Murer & Kinne, 1980). The maximum gradients of solute concentration formed in that way correspond to a factor of about 10-fold or

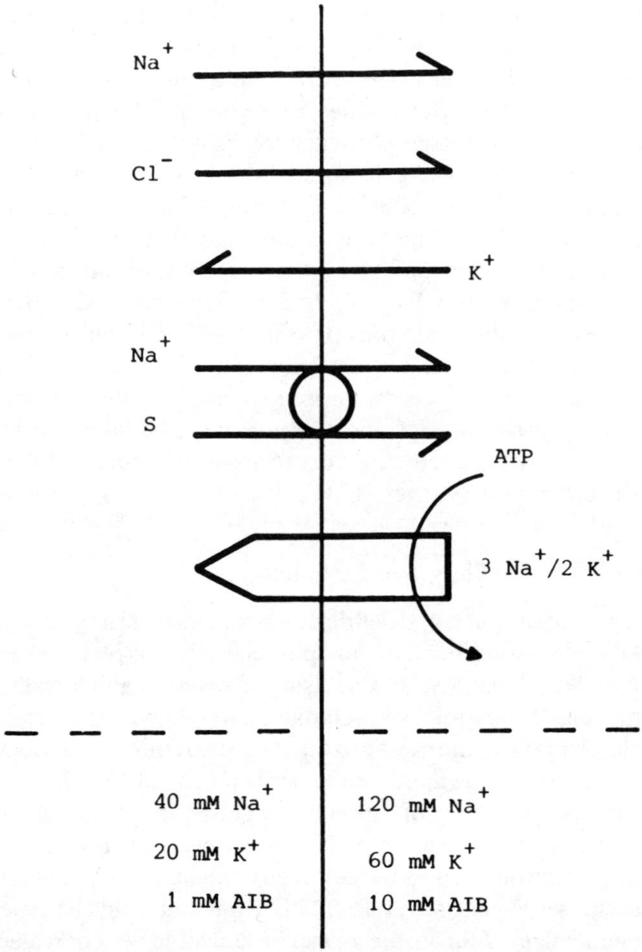

Fig. 2.9 Approximate ionic distributions that mouse ascites tumour cells maintain in the presence of selected amounts of nigericin when energy metabolism depends on lactate oxidation. The normal gradient of sodium concentration is inverted and yet a considerable accumulation of 2-aminoisobutyrate is maintained. The factors governing the membrane potential are not fully understood but may in these circumstances depend principally on the electrogenic action of the sodium pump overriding diffusion potentials due to K^+, Cl^- or Na^+ ions. (Taken with permission from Eddy, A. A. (1985). *Annals of the New York Academy of Science*, **456**, 57–62.)

6 kJ mol^{-1}. Analogous observations with isolated intestinal epithelial cells depleted of ATP, a system exhibiting a stoichiometry of two Na^+ per carbohydrate molecule, have shown that gradients of α-methylglucoside of up to 5.8kJ mol^{-1} form in these circumstances (see Kimmich & Randles, 1984). Furthermore, Turner and colleagues studied the way that renal vesicles, containing a higher glucose concentration than the surrounding medium ($\Delta\mu^S \simeq 4.5$kJ mol^{-1}), maintained that glucose concentration gradient when $\Delta\mu^{Na}$ was deliberately varied in a series of parallel assays (Figure 2.10; Table 2.3). They concluded that the later proximal tubule contains predominantly a symport mechanism translocating about two Na^+ ions and glucose, whereas the early proximal tubule contains predominantly a similar mechanism translocating one Na^+ ion and glucose. The larger Na^+ stoichiometry may facilitate the reabsorption of glucose from the urine (Fukuhara & Turner, 1984).

2.6 **Gradient coupling with organic anions**

Work with membrane vesicles derived from the brush border or the basal membranes of renal or gut epithelial cells has revealed a wide range of Na^+ dependent mechanisms involved in anion transport. Earlier electrophysiological studies had shown that whereas aspartate uptake in the newt proximal tubule required Na^+ ions, no net transfer of charge appeared to occur (Hoshi, 1976). In contrast Na^+ dependent uptake of aspartate or glutamate in the rat kidney was associated with a net uptake of positive charge (Fromter, 1979). (Distinct electroneutral and electrogenic modes of L-lactate uptake also occur; see Schell *et al.*, 1983.) This behaviour would be expected if one equivalent of Na^+ in the former instance and two equivalents in the latter instance accompanied each equivalent of aspartate across the luminal membrane. Study of glutamate uptake in rat or rabbit renal vesicles revealed unexpected complications. When K^+ was absent from the system glutamate uptake appeared to be an electroneutral process. However, in the presence of K^+ ions, the system appeared to be energised both by symport with Na^+ ions and antiport with K^+ ions and, at least in the rat preparations, then involved a net transfer of positive charge with the amino acid (Burckhardt *et al.*, 1980; Schneider & Sacktor, 1980 and Chapter 5).

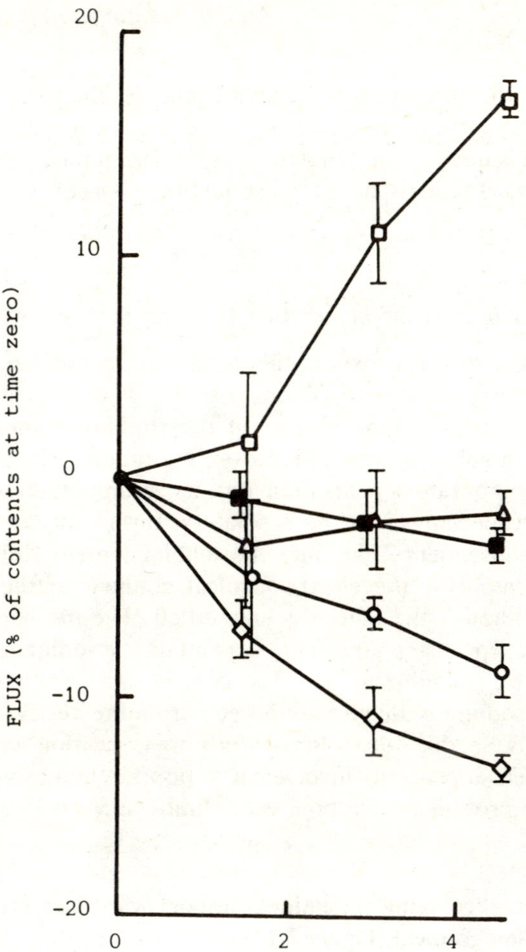

Fig. 2.10 Retention of glucose by brush border vesicles from the outer medulla, as a function of the membrane potential. The vesicles contain a concentration of glucose five times that in the surrounding medium but the same concentration of Na^+ ions as the latter solution. The membrane potential was poised at one of four selected values by controlling the concentration of K^+ ions outside the vesicles in the presence of valinomycin. Glucose retention was compared with the retention in a control lacking Na^+ ions. The conclusion drawn is that a membrane potential equivalent to three-fold concentration gradient of K^+ roughly balanced a five-fold concentration gradient of glucose. (Taken with permission from Fukuhara, Y. & Turner, R. J. (1984). *Biochimica et Biophysica Acta*, **770**, 73–8.)

Whether the presence of K^+ actually modifies the properties of the glutamate carrier, or whether two distinct carrier systems are involved is not known. Only the electroneutral mode of glutamate uptake was found in the rat intestinal brush border (Corcelli *et al.*, 1982).

2.7 Renal organic anion secretion driven by the sodium gradient

The ability of the proximal tubule of the mammalian kidney to secrete foreign organic anions has been widely studied and provides an interesting variant of the traditional mode of Na^+ dependent solute accumulation from the tubule. The secretion of *p*-aminohippurate appears to involve its accumulation by an active transport mechanism at the basolateral surface of the enterocyte and its subsequent efflux into the tubular lumen. Efflux may be driven simply by the electrochemical gradient of the hippurate anion, although the matter is not settled. The mechanism of the active transport step appears to depend on the sodium gradient, as basolateral membrane vesicles concentrate *p*-aminohippurate when a sodium gradient is imposed across the vesicle membrane (Moller & Sheikh, 1983). In constrast organic cation secretion into the lumen appears to involve an antiport relationship with the influx of protons at that boundary (Inui *et al.*, 1985).

2.8 Indirect coupling of peptide transport or cation secretion to the sodium gradient (Figure 2.11)

An alternative way of using the energy inherent in the sodium electrochemical gradient acting across the renal brush border resembles mechanisms found in bacterial systems (Eddy, 1978). The solute in question would be absorbed through a proton symport and the absorbed protons would then exchange with extracellular Na^+ through the Na^+/H^+ exchange mechanism that occurs in the brush border. The electrogenic renal absorption of certain dipeptides with protons as co-substrates may function in that manner (Takuwa *et al.*, 1985; compare Boyd & Ward, 1982) as may the absorption of certain di and tripeptides at the brush border of the rat intestine (Ganapathy *et al.*, 1985).

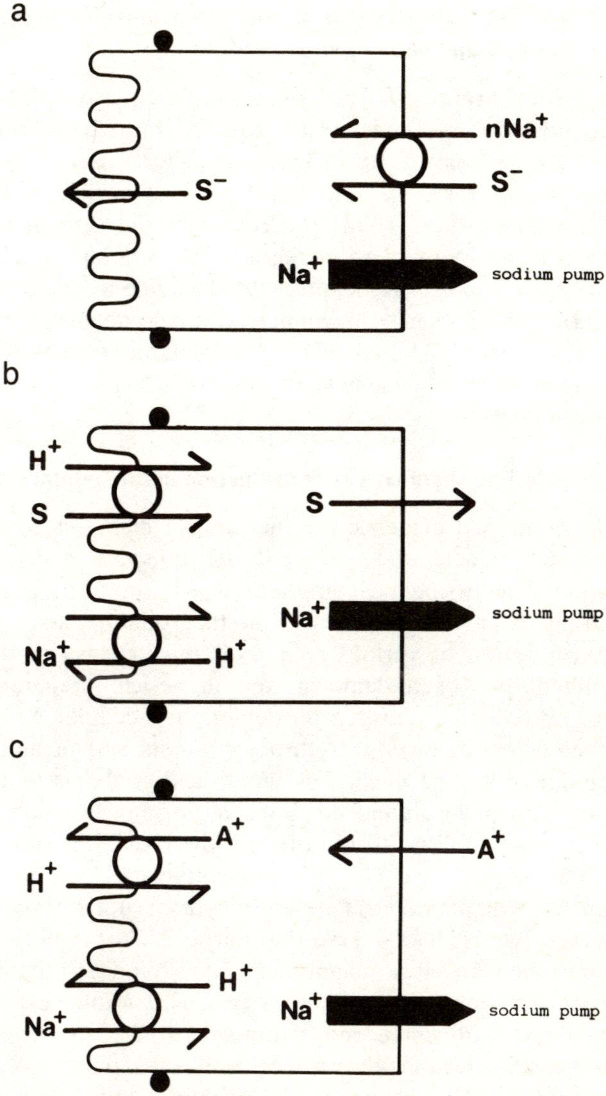

Fig. 2.11 Alternative ways of driving solute transport by the sodium gradient maintained by the sodium pump (see text). (*a*) Symport of an anion with *n* Na^+ ions at the basolateral surface and secretion of the anion at the apical surface. Mechanism of charge neutralisation not shown. (*b*) Symport of S with protons at the apical surface accompanied by Na^+/H^+ exchange, and efflux of S at the basolateral surface. (*c*) Cation absorption at the basolateral surface, followed by secretion in exchange with protons at the apical surface.

2.9 The link between solute accumulation and energy transduction through the sodium pump

As outlined on p. 59, when solutes are absorbed with Na^+ ions at the brush border, the latter are expelled through the sodium pump and can be assayed, for instance as a short circuit current. It has been known for some time that the rate of working of the sodium pump under these conditions may increase several fold in the absence of a permanent increase in the cellular concentration of Na^+ ions (Hudson & Schultz, 1984). While a volume-regulatory efflux of K^+ ions may play a part in the homoeostatic mechanism (but see Brown & Sepulveda 1985) the nature of the signal causing an increase in the activity of the sodium pump has yet to be determined.

2.10 Mechanism of energy transduction in the sodium symport

The combined evidence outlined above makes it very probable that amino acid and carbohydrate transport is driven by the sodium electrochemical gradient which, in turn, is established during the hydrolysis of ATP at the sodium pump. Significant concentration of various solutes at the expense of the sodium gradient has been demonstrated in vesicle preparations from various tissues and also, after depletion of cellular ATP, in intact enterocytes, in pigeon erythrocyte ghosts and in mouse ascites tumour cells. The maximum amino acid gradients that the latter system formed were indeed almost as large as those formed during energy metabolism. Study of the flow ratios has shown that, in certain tissues, one Na^+ ion per solute molecule serves as co-substrate in amino acid or carbohydrate uptake and, in other tissues, two Na^+ ions serve that purpose. Furthermore, comparison of the respective magnitudes of $\Delta\mu^{Na}$ and $\Delta\mu^{S}$ that the mouse ascites tumour cells maintain in various physiological conditions, indicates, with some reservations, that $\Delta\mu^{Na}$ is the sole and sufficient source of energy for the amino acid pump represented by System A. The comparable quantitative studies that have been done with renal vesicles and glucose also indicate that tight coupling is achieved at least in the Na^+-symport of the early proximal tubule. The systems translocating two Na^+ ions per glucose molecule deserve further study from that standpoint.

Because the protein components of these symports have yet to be characterised (Semenza *et al.*, 1984), one can only speculate about the molecular mechanisms underlying the different kinetic steps that are represented in Figure 2.2. In the most general terms coupling between their respective gradients is governed by the extent to which the substrate and co-substrate traverse the membrane together as opposed to separately. A relatively low but finite rate of solute uptake through the carrier in the absence of Na^+ ions and the ability of the latter to promote substrate binding through a random mechanism would reflect, on the one hand, a low probability of translocation of the binary species CS and, on the other hand, a conformational change associated with the binding of Na^+ ions that assists the subsequent binding of S. Efficient coupling also involves a low probability of translocation of the form NaC (see Eddy, 1968a and b; Heinz *et al.*, 1972). Because the translocation probabilities of the unloaded carrier C and the ternary form NaCS are necessarily both relatively large, each form probably representing a distinct conformation of the cis or trans facing carrier, it is tempting to view the binary forms as intermediate structures deformed by attachment of the single ligand. Recent attempts to detect the transition between the cis and trans conformations of a Na^+ independent glucose carrier (Appleman & Lienhard, 1985) and the conformational changes accompanying the attachment of Na^+ ions to the intestinal glucose symport (Peerce & Wright, 1985) are of great interest in this general context.

Another important set of possibilities concerns the locations of the sites binding the substrate and co-substrate when the carrier faces inwards or outwards. If these sites always lie within the plane of the bilayer, as has been suggested in a related context (Brooker & Wilson, 1985), the simple notion of the membrane potential affecting only the translocation steps in the carrier cycle (p. 62) may be implausible. An elegant formulation of some of these problems occurs in Lauger (1980).

2.11 **Further reading**

(Other literature is listed in the compiled References at the end of the book.)

Alvarado, F. & Van Os, C. H. (1986). *25 years of research on the brush border membrane and sodium-coupled transport.* INSERM Symposium Series vol. 26, Elsevier, Amsterdam.

Crane, R. K. (1977). The gradient hypothesis and other models of carrier-mediated active transport. *Reviews in Physiology, Biochemistry and Pharmacology*, **78**, 99–159.

Eddy, A. A. (1978). Proton-dependent solute transport in microorganisms. *Current Topics in Membranes and Transport*, **10**, 279–359.

Eddy, A. A. (1985). Sodium cotransport systems and the membrane potential difference. *Annals of the New York Academy of Sciences*, **456**, 51–62.

Heinz, E. (1978). *Mechanisms and energetics of biological transport.* Springer-Verlag, Berlin.

Stein, W. D. (1986). *Transport and diffusion across cell membranes.* Academic Press, New York.

Wright, J. K., Seckler, R. & Overath, P. (1986). Molecular aspects of sugar: ion co-transport. *Annual Review of Biochemistry*, **55**, 225–48.

3

Electrical changes evoked by ion cotransport with amino acids: a study in the pancreatic acinar cell

O. H. Petersen MRC Secretary Control Research Group
Department of Physiology University of Liverpool.

3.1 **Introduction**

The pancreatic acinar cells have a high rate of protein synthesis and therefore also accumulate amino acids with a high intensity (Schulz & Ullrich, 1979). Studies by Tyrakowski *et al.* (1978) have shown that in the presence of a Na^+ gradient L-alanine is taken up by plasma membrane vesicles against a concentration gradient. This can most easily be explained by an electrogenic Na^+–amino acid contransport (Philo & Eddy, 1978b). This chapter will deal with the electrical manifestations of Na^+–amino acid co-transport. Na^+–amino acid co-transport cannot, however, be discussed in isolation from other important electrogenic events in the plasma membrane. The electrogenic Na^+–K^+ pump creates the all important transmembrane Na^+ and K^+ gradients and K^+-selective membrane channels are responsible for the electrical membrane potential which provides an important part of the driving force for the Na^+–amino acid co-transport. Figure 3.1 shows the interrelationship between the three ion transporters of interest at the basolateral plasma membrane. There is considerable electrophysiological information available about each of these three transporters and they will be considered in turn. (The basolateral plasma membrane of epithelia of various organs, including pancreas, is also dealt with in Chapter 5.)

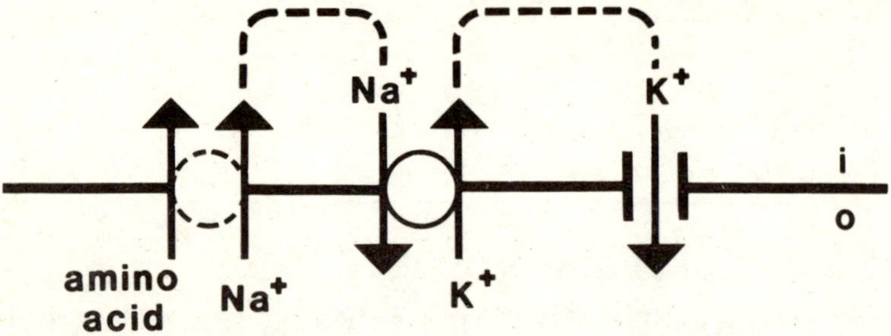

Fig. 3.1 A simplified scheme to account for the transport of amino acids and cations across the basolateral plasma membrane of the pancreatic acinar cell. The Na^+–amino acid co-transport system is coupled to the Na^+–K^+ pump which is closely linked to the K^+ channels. The inner and outer surfaces of the plasma membrane are denoted by i and o (see also Figure 2.2). Taken with permission from Singh, J. & Petersen, O.H. (1984). *Quarterly Journal of Experimental Physiology*, **697**, 531–40.

3.2 K^+ channels and the membrane potential

3.2.1 *K^+ channels in resting cell membrane*

The most direct method available for investigating ion channels in cell membranes is patch-clamp single-channel current recording. Instead of inserting a microelectrode into a cell a microelectrode tip is pressed onto the surface of a cell effectively isolating a small patch of membrane (1–10 μm^2). With appropriate low noise amplifiers the pico-ampere currents flowing through single channels when they open can be measured directly. The seal between the tip of a microelectrode and the outer surface of the cell membrane has under suitable conditions not only a high electrical resistance but is also mechanically very stable. The electrically isolated patch membrane can be pulled off the cell (excised) in such a way that the inside of the plasma membrane faces the bath solution (inside out) or alternatively so that the inside faces the solution in the micropipette (outside out). Figure 3.2 shows single-channel recordings from an excised inside-out membrane patch taken from the baso-lateral side of a pig pancreatic acinus. The membrane inside faces the bath which is filled with a K^+-rich (intracellular) solution whereas the exterior of the membrane faces the pipette solution which is Na^+-rich (extracellular). The intracellular K^+-rich solution has a very low ionised calcium concentration ([H $Ca^{2+}]_i$) with a value of 10^{-8}M in Figure 3.2*a* and 10^{-7}M in Figure 3.2*b*. At negative (normal) membrane potentials, short-lasting unitary upward current deflections are seen representing outward K^+ movement (from membrane inside to outside). When the membrane potential is made less negative, the amplitude of the single-channel currents increases due to the larger electrochemical gradient for K^+ movement into the pipette, and this is even more marked when the membrane potential is made positive. This is as predicted by Nernst-Planck electrodiffusion theory, but what could not have been predicted is that there are more and longer channel openings when the membrane potential is shifted from −30 mV to less negative values and that the channel stays open most of the time, except for very brief periods of closure, at the positive potentials. The probability of finding the channel in the open state (open-state probability p) therefore increases with decreasing negativity of the membrane potential. This K^+ channel is therefore voltage-gated. When $[Ca^{2+}]_i$ is increased 10-fold from 10^{-8}M (in *a*) to 10^{-7}M (in *b*) p markedly increases at the negative membrane potentials. The K^+ channel can therefore be controlled by changing the membrane potential and/or by changing $[Ca^{2+}]_i$. This type of channel is referred to as a Ca^{2+}- and voltage-activated K^+ channel and it is this channel that dominates the electrical characteristics of the basolateral plasma membrane in a large number of mammalian exocrine acinar cells (Petersen & Maruyama, 1984). In the normal resting acinar cell the open-state probability of the channel is low, but it is the only available exit pathway for K^+.

It is unfortunately very difficult to investigate single-channel currents in excised patches at physiological membrane potentials (i.e. about −60 mV) under conditions with quasi-physiological ion gradients because the driving force for K^+ outflow (the difference between the K^+ equilibrium (Nernst) potential and the membrane potential) is so small that the single-channel currents are barely detectable. However, when the K^+ concentration is raised to intracellular levels in the pipette and recordings are made from intact cells (as in the experiments represented by Figure 3.3) clear channel openings can be observed at the normal resting potential (pipette voltage, V_P = 0 mV) although the probability of channel opening is low. It is clear therefore that the Ca^{2+}- and volt-

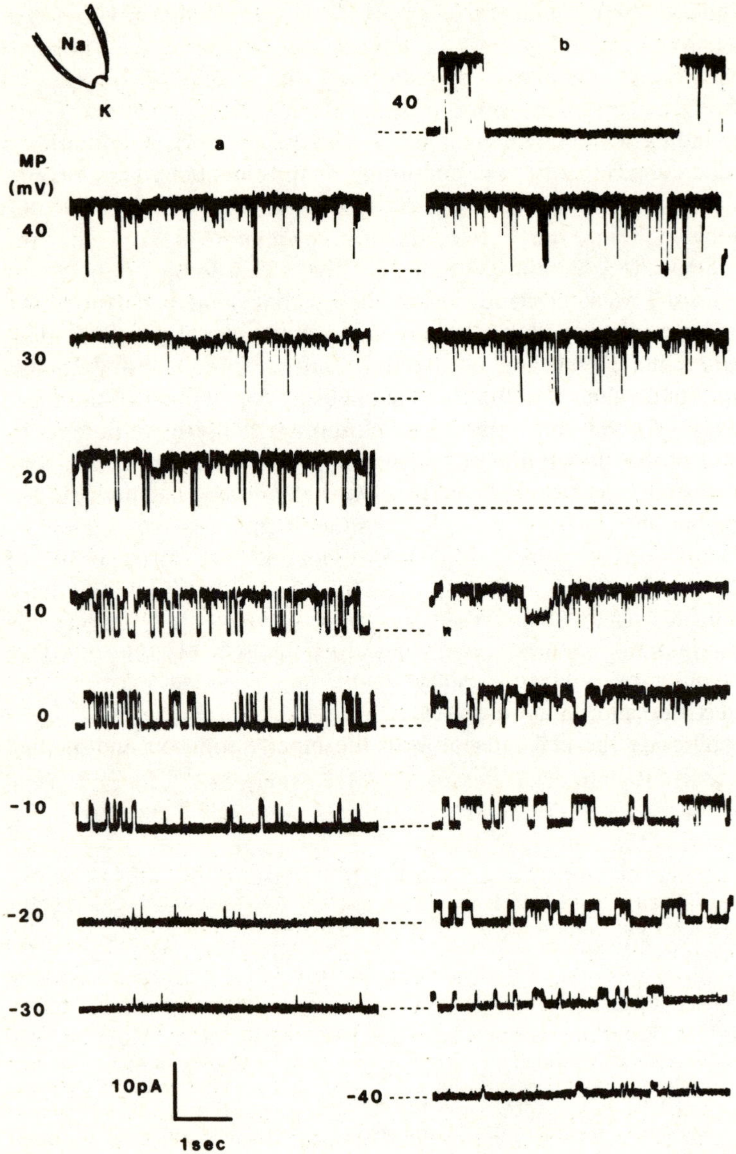

Na
K
a
b
MP
(mV)
40
30
20
10
0
-10
-20
-30
-40
10pA
1sec

age-activated K^+ channels must play a significant role in setting the resting potential. Ca^{2+}- and voltage-activated K^+ channels with the characteristics shown in Figure 3.2 and 3.3 have been directly observed in many cell types (Petersen & Maruyama, 1984) including pancreatic acinar cells (Maruyama *et al.*, 1983). These K^+ channels can be blocked by tetraethylammonium (TEA) acting specifically from the membrane outside (Iwatsuki & Petersen, 1985).

Single-channel recording is an attractive technique for characterising ion channels in the plasma membrane as it allows direct observation of the properties of individual molecules. The problem with this new technique is that it is difficult and certainly very time-consuming to obtain a quantitatively valid picture of the total ion conductance in the whole plasma membrane of a cell. A very useful extension of the patch-clamp technique for single-channel recording is the tight-seal whole-cell recording (Marty & Neher, 1983). Starting out from the recording situation shown in Figure 3.3 (cell-attached configuration) it is possible to break the isolated membrane patch and establish continuity between pipette interior and the cell interior. It has been shown that this allows a much better seal between pipette and cell membrane than can possibly be achieved with a conventional intracellular microelectrode and, in addition, there is the advantage of being able to equilibrate the cell interior with the pipette solution and in this way control the intracellular ionic composition.

Fig. 3.2 (opposite) Single-channel current recording from excised (inside out) pig pancreatic acinar membrane patch exposed (see cartoon) to extra cellular solution on the outside (pipette) and intracellular solution on the inside (bath). In the part of the experiment shown in (*a*) the ionised Ca^{2+} concentration in the bath $[Ca^{2+}]_i$ was 10^{-8}M. At the time when the traces shown in (*b*) were obtained from the same patch, $[Ca^{2+}]_i$ was adjusted to 10^{-7}M. The membrane potentials (MP) at which the individual traces were obtained is indicated. The current level when the channel is closed is indicated by horizontal dashed lines. Upward deflections represent outward current. Two traces represent the 10^{-7}M situation at a membrane potential of 40 mV. This is to indicate that while the channel was generally open there were times when the channel went from the open state into a state of prolonged closure. (Taken with permission from Maruyama, Y. *et al.* (1983). *Nature*, **305**, 228–32.)

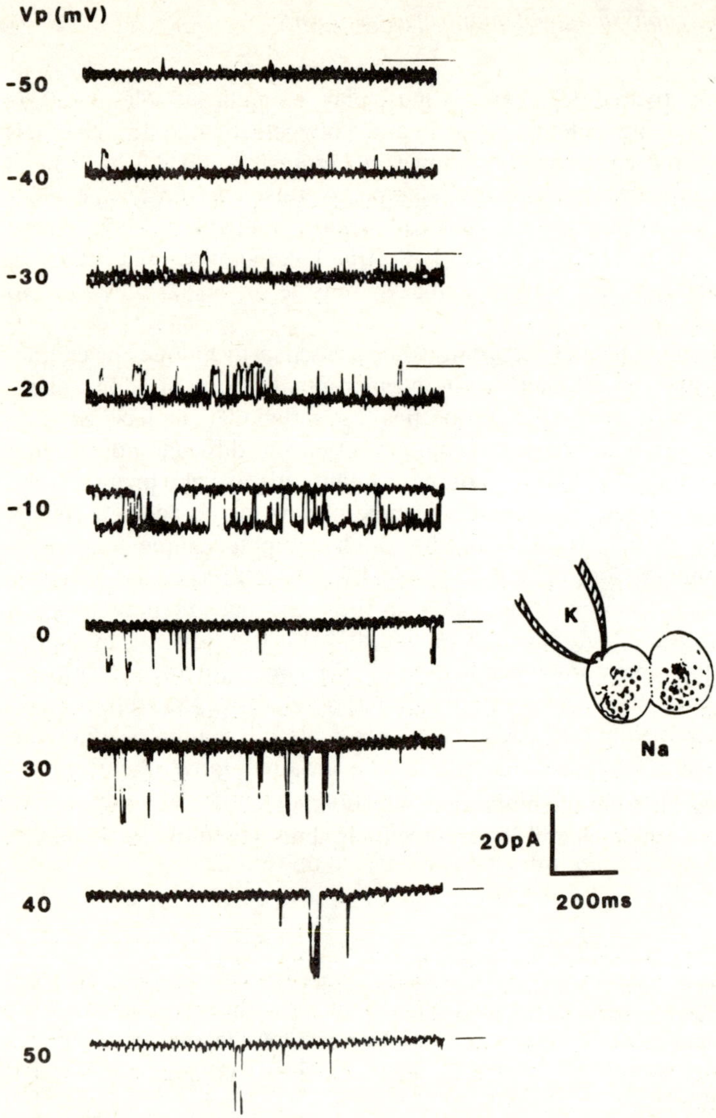

Fig. 3.3 Single-channel current recording from cell-attached membrane patch on isolated pig acinar cell. The cartoon shows the recording configuration. The recording pipette is filled with intracellular solution (high K^+ concentration) whereas the bath fluid is an extracellular solution (high Na^+ concentration). The traces show single-channel currents through the K^+ channel in the isolated patch area at different pipette potentials (V_p). (Taken with permission from Petersen, O. H. (1986), In: *The exocrine pancreas: Biology, pathobiology and disease* (V. L. W., Go, ed., pp. 141–61, Raven Press, New York.)

Figure 3.4 shows an example of the type of voltage-clamp experiment that can be carried out using the whole-cell recording configuration. The membrane potential is clamped at a holding level (−40 mV) and the currents associated with depolarising and hyperpolarising voltage jumps are recorded. As seen in Figure 3.4 the whole plasma membrane acts as a rectifier. Depolarising voltage jumps are associated with large outward currents whereas hyperpolarising jumps of the same magnitude generate very little inward current. Two factors contribute to this extreme rectifier property of the pig pancreatic acinar cell membrane. Firstly the K^+ channels, as already described, are very selective and as there is only a low K^+ concentration on the outside of the cell (4 mM) but a high concentration on the inside (150 mM) it is naturally much easier to encourage outward rather than inward K^+ movement. Secondly the K^+ channels as described above are voltage-gated. Depolarisation opens, whereas hyperpolarisation closes the channels. The K^+ conductance of the plasma membrane therefore acts as a very efficient machinery for preventing depolarisation, as a depolarising influence will activate the K^+ conductance and this (in a normal cell which does not of course operate under voltage-clamp conditions) evokes repolarisation. As seen in Figure 3.4, TEA specifically blocks the whole-cell K^+ currents from the membrane outside in agreement with the results from single-channel current experiments (Iwatsuki & Petersen, 1985).

3.2.2 *Computing the number of channels per cell*

The whole-cell voltage clamp currents can be regarded as the sum of many small elementary currents flowing through individual K^+ channels of the type characterised in the single-channel studies described above. There are a number of strong arguments which support this conclusion:

(i) The whole cell K^+ current is Ca^{2+}-activated and the Ca^{2+}-sensitivity is similar to that described for the single channels (Maruyama & Petersen, 1984, cited in Iwatsuki & Petersen, 1985.

(ii) The activation of the whole-cell currents occurs within the same voltage-range as described for single-channel currents (Maruyama & Petersen, 1984, cited in Iwatsuki & Petersen, 1985).

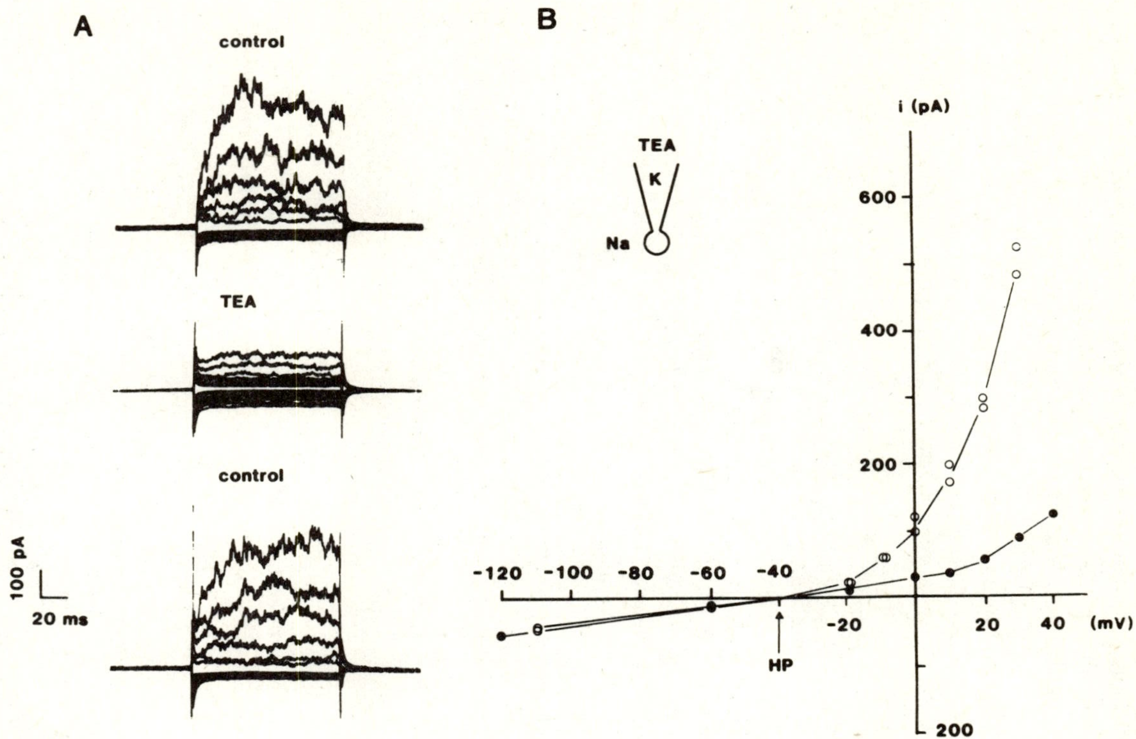

A
control
TEA
control
100 pA
20 ms
B
TEA
K
Na
i (pA)
600
400
200
200
-120
-100
-80
-60
-40
-20
20
40
(mV)
HP

(iii) Ca^{2+}-activated K currents can be relatively specifically blocked by low concentrations (about 2 mM) of TEA from the outside of the plasma membrane and TEA applied in this way blocks both single-channel and whole-cell K^+ currents in the pig acinar cells (Iwatsuki & Petersen, 1985).

If one accepts, on the basis of these three arguments that the whole cell current in the pig acinar cells is the sum of the elementary currents flowing through individual K^+ channels when they are open, it follows that the relationship between the whole cell current (I) and the single-channel current (i) is given by $I = Npi$, when N is the total number of operational K^+ channels and p the open state probability. At positive membrane potentials where p is close to 1 (Figure 3.2) then $N = I/i$.

In the resting pig acinar cells this approach has indicated that the total channel number is quite small, only about 50 channels per cell (Maruyama *et al.*, 1983). This small number of high-conductance channels would have caused problems if it had not been for the extensive cell-cell coupling in the acinar units. Several hundred cells form a functional unit (Petersen, 1976; 1980) and this means that on average several hundred channels will be open in the resting situation in one unit. With this number of channels the effect of openings and closings of single channels will be modest as the sudden closure of one channel, for example, would only reduce the overall conductance by about 1% whereas in a single cell this event might have caused a 100% loss of conductance. It would appear therefore that cell to cell junctional channels are essential for maintaining a stable resting potential in cells with small numbers of K^+ channels such as the pig pancreatic acinar

---

Fig. 3.4 (opposite) Effects of intracellular and extracellular TEA on whole-cell currents. Currents were evoked by depolarising or hyperpolarising voltage steps from a holding potential of −40 mV. Pipette (see cartoon) contained intracellular K^+-rich solution to which 2 mM TEA had been added. Bath was filled with extracellular Na^+-rich solution. (A) Traces of whole-cell currents before, during the presence of TEA (2 mM) in the bath, and after return to the control situation. Voltage steps of ±20 to 70 (80) mV were applied. (B) Whole-cell current-voltage relationship in the control situations (o), before and after TEA application, and in the presence of 2 mM TEA in the bath (•). HP = holding potential. (Taken with permission from Iwatsuki, N. & Petersen, O. H. (1985). *J. Memb. Biol.* **86**, 139–44.

cells. Many cells sharing their small number of K^+ channels can therefore overcome the threat of an unstable and unpredictable membrane potential arising from the stochastic nature of channel behaviour (Petersen, 1985).

3.3 The electrogenic Na^+–K^+ pump

The most dramatic demonstration of a direct pump contribution to the membrane potential can be achieved by maximally reactivating a previously blocked Na^+–K^+ pump. Experimentally, this can be achieved by exposing superfused pancreatic fragments to K^+-free solutions for more than 1 h. This arrests the Na^+–K^+ pump, since extracellular K^+ is required for operation; therefore, intracellular K^+ is lost and Na^+ accumulates. The resting potential becomes very low. Readmitting K^+ to the extracellular solution in the normal concentration (4.7 mM) results in an immediate marked hyperpolarisation, which is blocked by the specific inhibitor of the Na^+–K^+ pump, ouabain (see Petersen, 1973, in Petersen, 1980). It has further been shown that (i) ouabain added during the K^+ induced hyperpolarisation immediately abolishes the potential change; (ii) the K^+ induced hyperpolarization is not accompanied by a resistance change; (iii) a K^+ induced initial hyperpolarisation can be seen even when K^+ is admitted in such a high concentration that the K^+ equilibrium potential is less negative than the membrane potential at the time of readmission; and (iv) the K^+ induced hyperpolarisation occurs much more quickly than the re-establishment of a normal intracellular $[K^+]$. There is therefore abundant direct evidence for the electrogenic character of the Na^+–K^+ pump in mammalian gland cell membranes.

During normal resting conditions, however, the direct pump contribution to the resting potential is relatively small. Thus in both the pancreas and liver, ouabain application causes an immediate depolarisation of only a few millivolts. Thus it is only under conditions of extreme intracellular Na^+ loading that a direct pump contribution to the resting potential becomes dominant.

3.4 Electrical changes evoked by amino acids

Iwatsuki & Petersen (1980a and b) first demonstrated that neutral amino acids, such as L-alanine, when added to the bathing fluid of isolated pancreatic segments evoked marked membrane depolarisation and some increase in membrane conductance. Figure 3.5 shows a comparison of the effects of L-alanine and acetylcholine (ACh) on the membrane potential and input resistance of a mouse pancreatic acinar unit. The amino acid evokes a marked membrane depolarisation and a small decrease in input resistance whereas ACh evokes a much more dramatic resistance reduction. The depolarisation evoked by L-alanine is stereo-specific (Iwatsuki & Petersen, 1980a) (Figure 3.6) and is half-maximal at a concentration of 1.2 mM in the superfusion fluid (Iwatsuki & Petersen, 1980b). The amplitude of the L-alanine-evoked depolarisation is markedly reduced when the tissue is exposed to nominally Na^+ free solutions (Iwatsuki & Petersen, 1980b) and this effect is fully reversible (Figure 3.7).

The null potential for the action of several neutral amino acids was evaluated in experiments with two intracellular microelectrodes in which one electrode was used to set the resting potentials at varying levels by direct current injection and the other to record the membrane potential. Figure 3.8 summarises a series of experiments with L-alanine, L-valine and L-proline. It can be seen that the membrane potential at which the amino acids do not evoke any potential change is about +20 to +40 mV. These results indicate that when L-alanine is present in the extracellular fluid a Na^+ conductance pathway is opened and this result is the one expected from the known presence of the Na^+–L-alanine co-transport system.

In recent experiments it has been possible to study the L-alanine-evoked current under voltage-clamp conditions with control also over the intracellular ionic concentrations by using the patch-clamp method for whole-cell current recording from isolated cells (Marty & Neher, 1983). In recordings from single cells (or small aggregates of electrically coupled cells) alanine-induced inward currents of about 50 pA were observed for a single cell at 37°C at a membrane potential of 0 mV. Such L-alanine-evoked currents were abolished when Na^+ was replaced by K^+ or when the pH was lowered from 7.2 to 5.0. From the temperature dependence of the current an activation energy of 67 kJ/mol was estimated. The alanine-induced Na^+ currents were competitively inhibited by 2-(methylamino)-isobutyric acid indicating that L-alanine is transported by the A system (see Chapter 1, Table 1.1). By internal perfusion of the

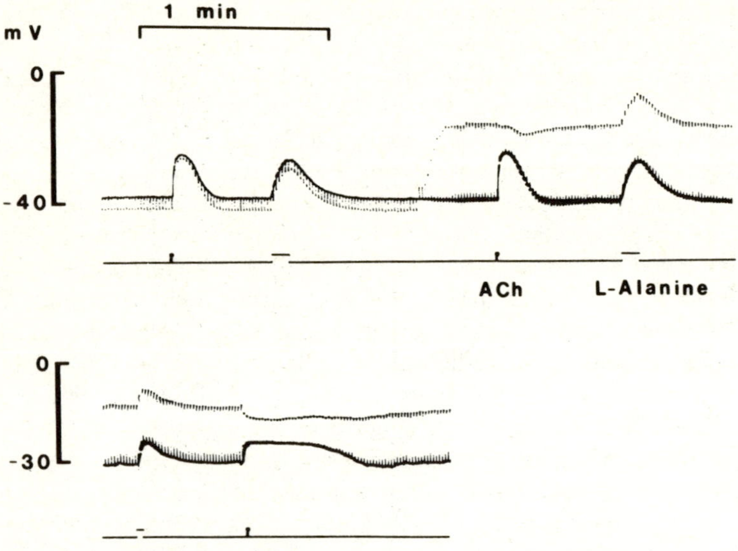

Fig. 3.5 Comparison of the membrane effects of L-alanine with ACh or gastrin in the same cells. In the upper trace ACh and L-alanine applied from two separate micropipettes by ionophoresis (80 nA, 0.5 s and 800 nA, 5 s, respectively) evoked depolarisations of similar size. During the first ACh and L-alanine stimulation, small hyperpolarising current pulses were repetitively injected through a microelectrode inserted into a neighbouring cell. Thereafter, the polarity of the injected current pulses was switched and the intensity increased (to 7 nA, 100 ms). In this way, the action of ACh and L-alanine could be simultaneously assessed at the spontaneous resting potential (heavy trace) (about −40mV) and the artificially lowered potential (interrupted trace) (about −14 mV). The reduction in resting potential from −40 to −14 mV just reversed the polarity of the ACh-evoked membrane change, whereas the L-alanine-evoked depolarisation was only slightly reduced. The lower trace shows a similar comparison between L-alanine and gastrin actions in another cell. Gastrin was added directly (3.5 nmol) to the bath. At −30 mV, both substances evoked depolarisation. At −13 mV, the gastrin response was reversed whereas the L-alanine response was only slightly reduced in amplitude. (Taken with permission from Iwatsuki, N. & Petersen, O. H. (1980*a*). *Nature*, **283**, 492–4.)

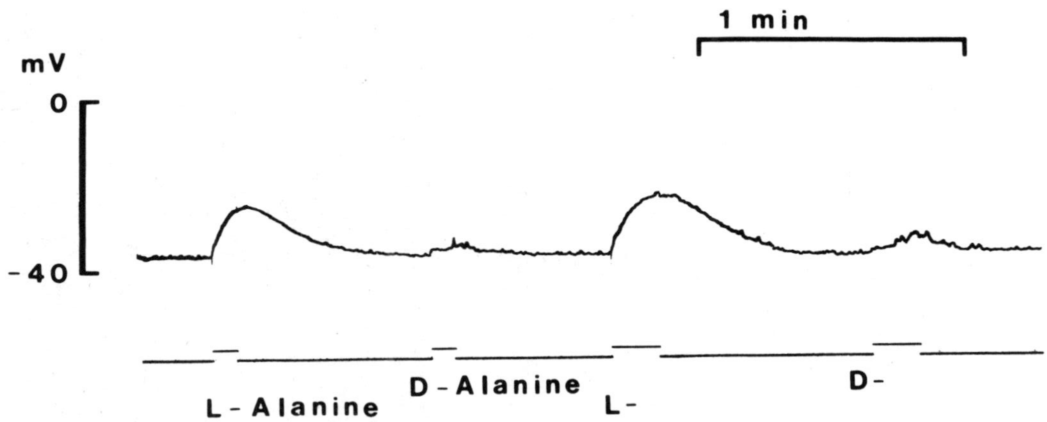

Fig. 3.6 Effects of ionophoretic L-alanine and D-alanine applications on the membrane potential in the same acinar cell (atropine 10^{-6}M present). Micropipettes were filled with either 1 M L- or D-alanine dissolved in 0.9% NaCl (pH 4.0). The marker signals in the bottom trace indicate periods of L- or D-alanine ejection near to the surface of the impaled acinar cell, using positive ejecting currents of 100 nA and 5 s (short marker signals) or 10 s (longer marker signal) duration. No retaining current was used. (Taken with permission from Iwatsuki, N. & Petersen, O. H. (1980*a*). *Nature*, **283**, 492–4.)

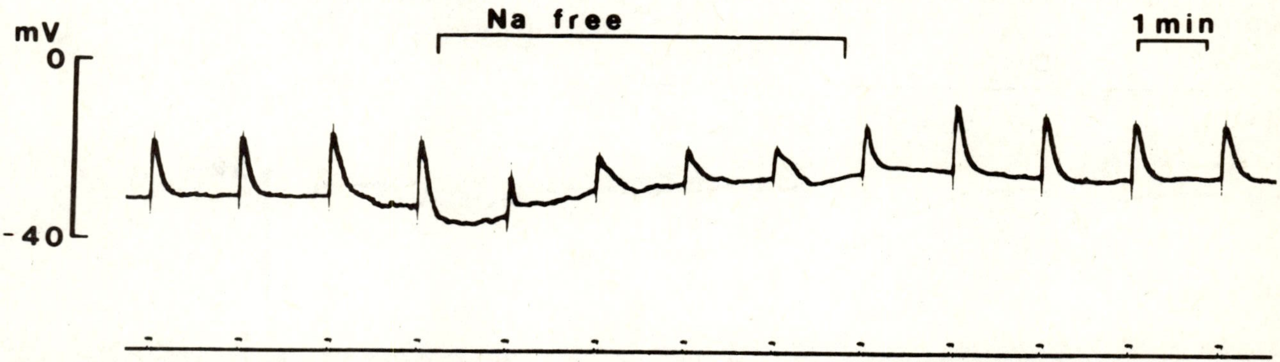

Fig. 3.7 The effect of replacing superfusion fluid Na^+ by $Tris^+$ on depolarisations evoked by micro-ionophoretic L-alanine application (100 nA, 2 s). (Taken with permission from Iwatsuki, N. & Petersen, O H. (1980*b*). *Pflügers Archiv*, **386**, 153–9.)

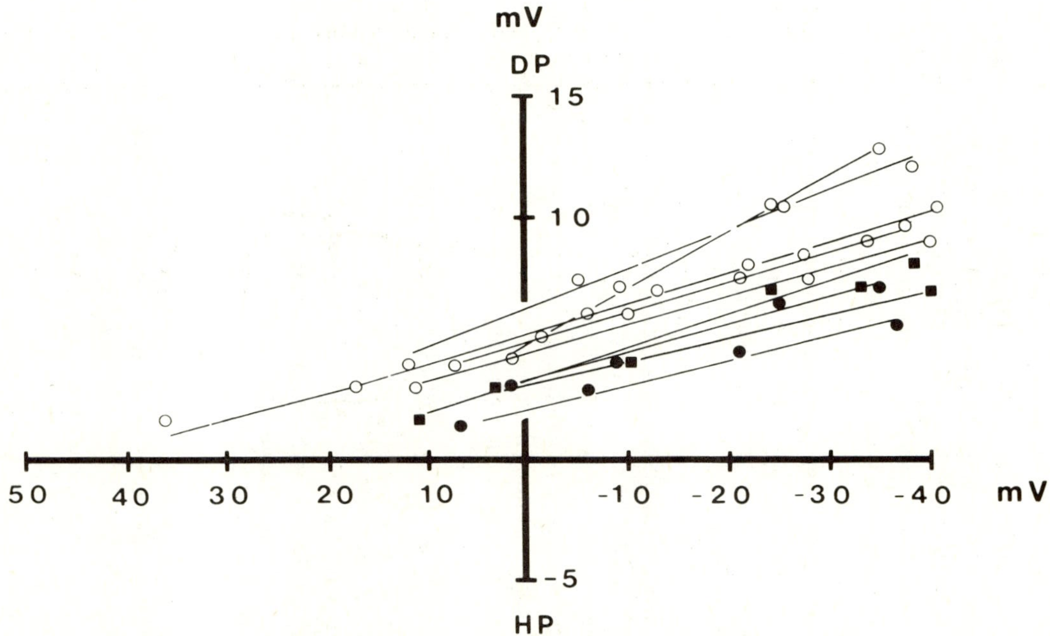

Fig. 3.8 The effect of varying the membrane potential on the amplitude of depolarisations induced by L-alanine (o), L-valine (●) and L-proline (■). The membrane potential was changed by injecting direct current through a second microelectrode inserted into a neighbouring coupled cell. Points connected by straight lines are from the same experiment on the same acinus. DP = depolarisation; HP = hyperpolarisation. (Taken with permission from Iwatsuki, N. & Petersen, O.H. (1980*b*). *Pflügers Archiv*, **386**, 153–9.)

patch pipette, the intracellular medium could be exchanged during the whole-cell recording experiment. When alanine was introduced into the cell by perfusion (or by preloading), large outward currents were observed at zero external alanine concentration. The amplitude of the outward current was comparable to the amplitude of the normally observed inward current (Jauch, *et al.*, in Alvarado & Van Os, 1986).

'Zero-trans' conditions (finite concentrations on one side and vanishing concentrations on the other) can be useful in order to evaluate kinetic parameters of transport systems. In this type of experiment there would, for example, initially be a Na^+- and alanine-free solution inside and outside the cell. Exchanging the outside solution with one containing 100 mM Na^+ only evokes a very small inward current (at a membrane potential of 0 mV) in spite of the enormous inwardly-directed Na^+ gradient, but when subsequently L-alanine is added to the outside solution, still at the same Na^+ concentration, there is a large inward current demonstrating the tight coupling between Na^+ and amino acid transport. In such 'zero-trans' experiments the magnitudes of the peak inward currents as functions of extracellular L-alanine and Na^+ concentrations have been investigated. The peak inward currents can be represented by Michaelis-Menten equations. The half-saturation concentration of alanine at the external site decreased from 18 mM at $[Na^+]_0$ = 5 mM to 2.9 mM at $[Na^+]_0$ = 150 mM. The half-saturation concentration of Na^+ was found to decrease from 63 mM at $[\text{L-ala}]_0$ = 1 mM to 14 mM at $[\text{L-ala}]_0$ = 10 mM (Jauch, Petersen & Läuger, 1986). From reversal potential measurements in the presence of various alanine and Na^+ concentration gradients a Na^+/alanine stoichiometric ratio of 1:1 could be inferred (Jauch, Petersen & Läuger, 1986).

All the neutral amino acids evoke similar electrical effects and the null potentials for the actions of L-alanine, L-phenylalanine, L-valine, L-proline and glycine are the same (Iwatsuki & Petersen, 1980a and b; Laugier & Petersen, 1981). Interaction experiments demonstrate mutual inhibition of depolarisation responses. It can be shown, for example, that the depolarising action of L-proline is much reduced in the presence of L-alanine even when the depolarising effect of L-alanine has been cancelled by direct injection of hyperpolarising current (Figure 3.9).

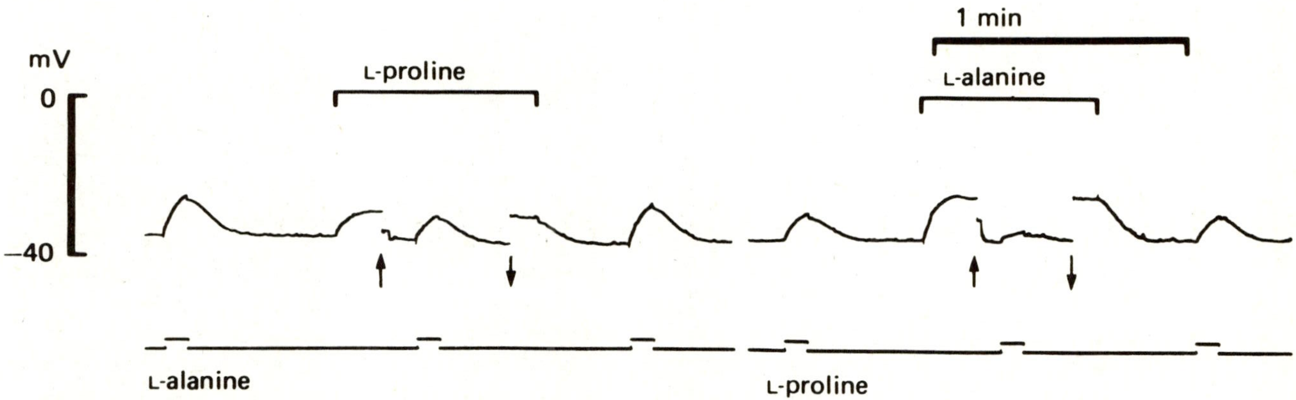

Fig. 3.9 Effects of one amino acid alone or in the presence of another amino acid compared at the same level of membrane potential. The protocol in the two experiments are similar. First a 500 nA, 5 s ionophoretic pulse of, e.g. L-alanine, is applied evoking the usual depolarisation. After recovery continuous L-proline (e.g.) stimulation is carried out (500 nA ejecting current from a second extracellular micropipette). When a steady depolarisation has been achieved direct current is injected through a second intracellular microelectrode (↑) to bring back the potential to the normal resting level. Now a short pulse of the amino acid first applied (e.g. L-alanine) is repeated. After recovery from a relatively smaller depolarisation the direct current injection is discontinued (↓) and the L-proline stimulation is subsequently stopped. Finally yet another short pulse stimulation with the first applied amino acid (e.g. L-alanine) is carried out. Two interaction experiments for L-alanine and L-proline are shown. (Taken with permission from Iwatsuki, N. & Petersen, O. H. (1980*b*). Pflügers Archiv, **386**, 153–9.)

The action of the basic amino acid L-arginine is different from that of the neutral ones. The maximal depolarisation evoked by supramaximal L-arginine concentrations is much smaller (about 4 mV) than that observed for L-alanine (about 18 mV) and $E_{L\text{-}arg}$ is much less positive than $E_{L\text{-}ala}$. In one series of experiments in which $E_{L\text{-}ala}$ and $E_{L\text{-}arg}$ were both measured with the same electrodes in the same acinar units the values were +30 mV for L-alanine and −7 mV for L-arginine (Laugier & Petersen, 1981). L-arginine is also unable to inhibit L-alanine-evoked depolarisations and similarly L-alanine fails to reduce L-arginine-evoked membrane depolarisation, although in the same experiments the already described mutual inhibition between the different neutral amino acids can be readily demonstrated (Laugier & Petersen, 1981). While it is clear that L-arginine opens up a membrane pathway with a different permeability from the one activated by the neutral amino acids there is no positive evidence yet available pointing to the ionic mechanisms underlying the arginine-evoked depolarisation. In addition to the amino acids already mentioned it has been shown that a number of other amino acids such as the neutral asparagine, glutamine, hydroxyproline, isoleucine, leucine, serine, tryptophan, tyrosine and methionine, the basic histidine, lysine and ornithine and the acid aspartic acid, glutamic acid, cysteine and cystine all evoke membrane depolarisation in varying degrees. Detailed dose-response curves and ionic mechanisms have, however, not yet been investigated.

3.5 **Conclusion**

In order to manufacture and secrete protein it is necessary for the acinar cells to have well-developed Na^+ dependent amino-acid uptake systems and in order for these systems to work both the Na^+–K^+ pump and the resting K^+ channels are essential since the resting potential is providing a major driving force for the 'downhill' Na^+ uptake linked to the 'uphill' amino acid transport. A schematic drawing of the plasma membrane with the three essential transport proteins is shown in Figure 3.1. The Na^+–K^+ pump creates the inwardly directed Na^+ gradient which is utilised to move neutral amino acids into the cell. Operation of the Na^+–amino acid co-transporter depolarises the cell and tends to increase $[Na^+]_i$. The depolarisation turns on voltage-activated K^+ channels, in those cells where they are present, and this partially repolarises the cell and helps to maintain the driving force for Na^+–amino acid transport. The increase in $[Na]_i$ activates the Na^+–K^+ pump and the K^+ that is accumulated when Na^+ is extruded exits via the K^+ channel.

3.6 **Further reading**

(Other literature is listed in the compiled References at the end of the book.)

Hille, B. (1984). *Ionic channels of excitable membranes.* Sinauer, Sunderland, Mass.

Iwatsuki, N. & Petersen, O. H. (1980). Amino acid-evoked membrane potential and resistance changes in pancreatic acinar cells. *Pflüger's Archiv*, **386**, 153–9.

Iwatsuki, N. & Petersen, O. H. (1980). Amino acids evoked short-latency membrane conductance increase in pancreatic acinar cells. *Nature*, **283**, 492–4.

Jauch, P., Petersen, O. H. & Läuger, P. (1986). Electrogenic properties of the sodiumalanine cotransporter in pancreatic acinar cells. I. Tight-seal whole-cell recordings. *J. Membr. Biol.* **94**, 99–115.

Katz, B. (1966). *Nerve, muscle and synapse.* McGraw-Hill, New York.

Petersen, O. H. (1976). Electrophysiology of mammalian gland cells. *Physiological Reviews*, **56**, 535–77.

Petersen, O. H. (1980). *Electrophysiology of gland cells.* Academic Press, London.

Petersen, O. H. & Maruyama, Y. (1984). Calcium-activated potassium channels and their role in secretion. *Nature*, 307, 693–6.

Petersen, O. H. & Petersen, C. C. H. (1986). The patch-clamp technique: recording ionic currents through single pores in the cell membrane. *News in Physiological Sciences*, **1**, 5–8.

Singh, J. (1984). L-alanine and L-phenylalanine activate Na^+ and K^+ conductance pathways in the exocrine mouse pancreas. *Pflüger's Archiv*, **402**, 176–84.

4

Amino acid transport systems in mammalian red cells

J. C. Ellory University Laboratory of Physiology University of Oxford

4.1 **Introduction**

The red cell is one of the classical isolated cell systems which has been used for membrane transport studies. It is therefore surprising that relatively little attention has been paid to amino acid transport in erythrocytes, at least compared with the growth of literature on the sodium pump, glucose transport system, capnophorin (the Band 3 anion transporter) and even nucleoside and choline transporters (see Ellory & Lew, 1977). A number of workers have recently been attracted to the red cell as a model system, since it is possible to measure fluxes under defined intra- and extracellular conditions, and to obtain precise kinetic data.

It is perhaps relevant to consider why red cells need amino acid transport systems. As we have argued in the past (Young & Ellory, 1977), there are several obvious reasons. Glutathione (GSH) biosynthesis, which is active in red cells requires adequate supplies of the three precursor amino acids, cysteine, glycine and glutamate, which immediately suggests transport systems of widely different substrate specificity (Ellory *et al.*, 1983 – see Bannai & Tateichi, 1986). Other possibilities include:

(i) the presence of the amino acid transport systems as a functional relic of the large requirement for protein synthesis in the reticulocyte or earlier red cell precursor (see Chapter 1, section 1.2.1.);
(ii) the controversial idea of Elwyn (1966) that the red cell plays a role in interorgan transport of amino acids, a notion which has recently been revived and critically reviewed (Christensen, 1982);

(iii) amino acids can also be 'accidental' substrates of other transport systems – as seen for the transport of glycine by Band 3 (Young *et al.*, 1981, and see below), but there may be other cases (the choline transporter is one obvious candidate).

Another recent and controversial finding is the apparent definition of a peptide transporter in the red cell membrane (King & Kuhel, 1984, 1985). This leads to the radical notion of amino acid transport in red cells for exporting the products of digested peptides.

It is the intention of this chapter to concentrate on mammalian erythrocytes, since they are simpler systems than the red cells of fish and birds. However it must be admitted that the transport of glycine, glutamate and taurine is important in the context of volume regulation in the nucleated erythrocytes of these latter species, and the transport capacities of some of these systems is very impressive (see for example Vidaver, 1964).

4.2 **Analysis of transport systems**

A major problem in studying amino acid transport systems lies in defining their specificities (Figure 4.1, see also Table 1.1). We do not have inhibitors such as ouabain for the sodium pump, to dissect the various systems. The main inhibitor studies to date have relied on non-specific sulphydryl group reagents such as PCMBS and NEM. Although these reagents have shown differential sensitivity in terms of selective inhibition of System ASC *vs* System L, they do not provide a basis for selecting definitively between any of the systems. The approach pursued by Christensen (see Chapter 1) in attempting to define specificity was the design of ideal model substrates, and the use of competitive inhibition. This strategy can be extended to accelerative exchange experiments, where the presence of trans-stimulation (e.g. of cysteine efflux by external L- but not D-alanine) can be used to indicate an affinity for the transport system in question.

4.2.1 *Genetic variants*

An increasingly successful alternative approach is the use of inter-species, or intraspecific variation for differentiating amino acid transport systems. We originally showed that a red cell mutation in sheep with genetically determined low GSH levels was associated with the absence of the transport System C (see Table 1.1) from the red cells (Young *et al.*, 1975). Subsequently, three separate variants of amino acid transport have been defined in the red cells of horses (Fincham *et al.*, 1985, 1987), and a further example of excess system glu transport in dog red cells has been documented (Maede *et al.*, 1982). One of the attractions of such variants is the possibility of making biochemical comparisons on cells differing in only one specific membrane transport system, offering exciting possibilities in terms of antibody production, and membrane fractionation. There are also large interspecific differences, which may be valuable in comparing amino acid transport where multiple pathways exist (see 4.4.3. below).

4.2.2 *Nomenclature of transport systems*

One aspect of amino acid transport which has recently received systematic attention is nomenclature, and the co-identity of amino acid transport systems from different cells, tissues and species. In this review on red cells I have attempted as far as possible to follow the outlines agreed recently (Bannai *et al.*, 1984), but there are inevitable semantic difficulties. In particular, there exist systems with almost identical substrate preferences (e.g. systems ASC, asc and C) but widely variant relative affinities, and differences in whether the System is Na^+ dependent or not. It will remain to be shown at the level of protein structure of the isolated transporters exactly how different these systems really are. Figure 4.1 summarises the main mammalian amino acid transport systems as known today for red cells. Each of these will be discussed briefly in turn below, with the evidence for their specificity, and separate existence. It is however appropriate initially and briefly to discuss techniques, since this is crucial to understanding the limitations in defining amino acid transport systems.

4.3 **Methodology and kinetic analysis**

The fact that red cells represent a homogeneous population of cells, without any extracellular compartment, makes them ideal for studying amino acid transport with radioactive tracer techniques. In fact we have reviewed the methods commonly employed in tracer experiments recently, and so do not need to go into fluxes in detail here (Young & Ellory in Ellory & Young, 1982). It is sufficient to mention some of the possible limitations and problems with this technique. Although red cells are a very pure cellular preparation by most physiological criteria, they do have some contamination with white cells and platelets which has led to biochemical problems in the past (e.g. the presence of glutaminase). Filtration through boiled cotton wool has helped eliminate this problem, as has the detailed determination of time-courses of amino acid fluxes, which can sometimes identify multiple compartments with different rates of transport and equilibrium spaces. A related problem for tracer fluxes concerns the purity of the radioactive material. This is of concern both initially, where some tritiated amino acids are notoriously unstable (e.g. tryptophan), and experimentally, where biochemical modification of the amino acid during the course of an experiment should be considered (particularly with pre-loading and efflux experiments) since metabolism may alter the apparent concentration of intracellular amino acid. In this context, the recent increasing use of nuclear magnetic resonance (NMR) and high pressure liquid chromatography (HPLC) techniques to identify specific amino acids chemically is promising (see Barret, 1985). Although there are significant problems with both methods, such as the high hematocrit required to see an adequate NMR signal, and the problems of resolving certain components by HPLC, it is clear that more data will be obtained by these methods in the future. However, the great attraction of measuring tracer fluxes by the microcentrifuge oil-separation or multiple washing techniques lies in the flexibility of the method (times from 5 s to 1 h or longer are used routinely), and the large number of replicates it is possible to process in a short time. This has allowed relatively high quality kinetic data to be achieved easily.

Although not strictly methodology, it is perhaps appropriate to consider kinetic analysis of amino acid transport systems at this point. This area is dominated by the treatment of Lieb & Stein, perhaps summarised most clearly recently by Lieb (in Ellory & Young 1982). Although our views of carriers have changed in terms of gating and reorientation of membrane-embedded protein units rather than any transmembrane diffusion or movement implicit in early ideas, the relatively simple formulations of Lieb allow considerable mechanistic information to be deduced from kinetic experiments. The human red cell L System, with leucine as substrate has been subject to at least two major kinetic analyses (Hoare, 1972; Rosenberg, 1981a), and Rosenberg has applied a similar analysis to the T System (Rosenberg, 1981b). Recently, Fincham *et al.* (1985) have started analysing the asc system in horse red cells in the same kinetic way. The principal conclusions of Rosenberg's work, aside from the rejection of a simple pore hypothesis for leucine transport, are that the System behaves consistently with a simple carrier model. The data indicate that the dissociation and translocation of the carrier-substrate complex are both faster than the translocation of the empty carrier, and that no translocation step is rate determining. At equilibrium, the carrier complex is equally distributed across the membrane. This kind of analysis produces useful predictions on the functioning of the transport protein, and will be applied increasingly to other systems in the future.*

4.4 **Amino acid transport systems in red cells**

All amino acid transport systems so far described for mammalian red cells are summarised diagrammatically in Figure 4.1. The areas

---

*Separate kinetic characterisation of entry and exit of Na-coupled glycine transport was reported by Wagensberg & Blostein (1985). They showed that transport appeared to be symmetrical for both glycine and sodium fluxes. These results were obtained in membrane vesicles prepared from sheep reticulocytes by means of a method first employed in human red blood cells (Steck, 1974 – see Wagensberg & Blostein, 1985) which allow the preparation of vesicles of distinct sidedness, i.e. either inside-out (cytosolic side) or right-side-out. The asymmetry of inward (cytosolic) and outward (extracellular) facing binding sites of the glucose carrier has been demonstrated in the erythrocyte, e.g. for the glucose carrier see Widdas (1980), for the choline carrier see Deves & Krupka (1984). Widdas's review is of general interest as it discusses the structural differences affecting both lipids and proteins in the plasma membrane, and also the physiological implications of asymmetric transport (the natural asymmetry of the composition of the cytoplasm and the extracellular medium is obvious). (Editor's note.)

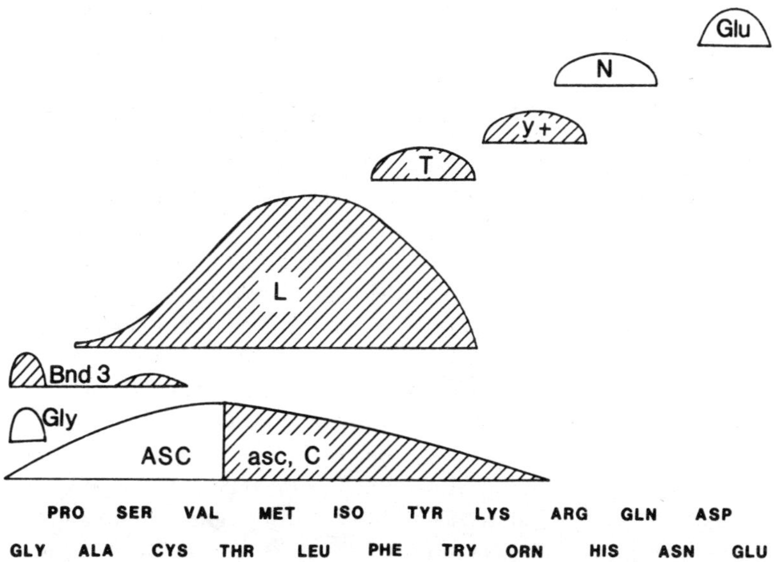

Fig. 4.1 Diagrammatic representation of all currently known amino acid transport systems in mammalian red cells. The area and shape corresponding to each transporter has been deliberately smoothed to present an overall schematic picture. No conclusions should be drawn from the apparent size of each envelope, except to say that the two largest, L and ASC/asc/C are the most important in the red cells of most species. Hatched areas imply the system is Na^+ independent. For specific details of each system consult the following references: Band 3, Young *et al.* (1981); gly, Ellory *et al.* (1981); ASC, Al-Saleh & Wheeler (1982); asc, Fincham *et al.* (1985); C, Young *et al.* (1980); L, Al-Saleh & Wheeler (1982) and Rosenberg (1982a); T, Rosenberg *et al.* (1980); y^+, Gardner & Levy – see Young *et al.* (1980); N, Ellory & Osotemehin (1983); glu, Ellory *et al.* (1982).

indicating the substrate specificity of each system are deliberately represented as smooth shapes rather than discontinuous histograms, as is more often done, to put together results from more than one study, where appropriate, and to overcome the problem of how to deal with substrates which have a higher affinity, but lower V_{max} than others. Again, the size of the areas is not proportional to transport capacity, but systems L and ASC are intended to be represented as the principal transporters. In fact, it is apparent that the systems divide into six relatively narrow systems designed to carry a discrete group of amino acids, and two broad systems with wide substrate specificities. The presence of a system apparently designed to have the substrate specificity for transporting a particular amino acid does not necessarily imply that it will be transported on that system to a major, or even a significant extent under certain defined conditions.

This is emphasised by the histograms illustrated in Figure 4.2 where the combined pathways which account for the transport of three representative amino acids, tryptophan, glutamine and glycine are individually identified for uptake at 0.1–0.4 mM plasma concentrations. Thus although the T System is apparently designed for transporting tryptophan, 30% of the uptake in the case illustrated is by other routes. Glycine represents an even more complex example, since there are five distinct participating pathways. One immediate consequence of such a situation is the difficulty of carrying out kinetic studies on a particular transport system, since in the absence of definitive specific inhibitors, complete separation by competitive inhibition studies is often difficult to achieve. Another difficulty in interpreting uptake studies is deciding whether there are any exchange components in operation, or whether the cell is truly being studied in a zero-trans state. With washed human red cells it is likely that the internal amino acid concentrations are truly low, but during competitive inhibition studies, a substrate such as leucine can enter via System L very rapidly. Then if the internal leucine concentration is sufficiently high to promote exchange via System ASC for another substrate, misleading results in terms of the apparent effects of leucine as an external inhibitor of System ASC can occur. However if such caveats are perceived initially, it is possible to define the substrate specificity of the transport system with reasonable certainty.

The individual systems will now be reviewed with their particular individual properties being described in a little detail. Table 1.1 in Chapter 1 can also be consulted for particular details of these systems in other cell types.

4.4.1 *Neutral amino acids: system gly*

There are two dedicated systems for glycine transport identified in Figure 4.1: system gly, and Band 3. The gly system is unusual in

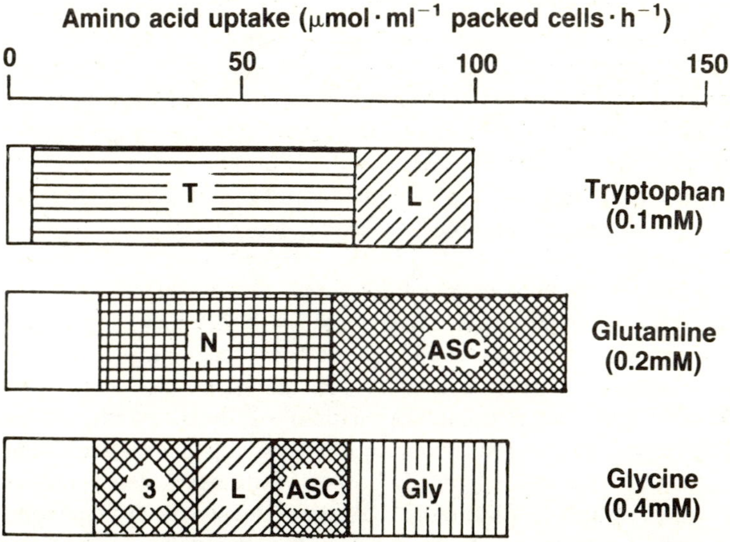

Fig. 4.2 Histogram illustrating the multiple transport systems for amino acid uptake by human red cells (at 37°C) for the substrates tryptophan, glutamine and glycine. Unshaded areas represent diffusion, or transport by uncharacterised routes. For definition of Systems see Figure 4.1. and text, and also Table 1.1. (Data taken from references in Figure 4.1.)

its very limited substrate specificity, carrying only glycine and sarcosine, and unique in showing an absolute requirement for Cl^- ions. This system was first characterised in avian erythrocytes (Vidaver & Shepherd, 1968), where it is important. It is tempting to speculate that the chloride-dependency may also be a parallel with the role of glycine as a neurotransmitter in opening chloride channels. Another role for glycine in certain animals, particularly marine invertebrates, is in volume regulation with the amino acid making a significant contribution to intracellular osmolarity. It is therefore also worth speculating whether there may be similarities between the Na, K, Cl co-transporter seen in many cell types and responsible for volume regulatory fluxes for example in avian red cells, and system gly (see Chipperfield & Perring, 1980), although it is certain that the two systems are not identical (Ellory *et al.*, 1981). As indicated in Figure 4.2, system gly plays a significant, but not dominant role in glycine transport under physiological plasma concentrations. Another significant contribution is made by capnophorin (Band 3) where we have indicated that glycine is an 'accidental' substrate. In fact, glycine is probably carried 10^7 times less on Band 3 than chloride ions, but nevertheless the flux is large enough to be important because of the relative abundance of Band 3 as a membrane protein (10^6 copies per red cell). Other amino acids transported include cysteine and serine, but not proline or any of the larger, or acidic amino acids.

4.4.2 *Neutral amino acids: System ASC*

Working from left to right on Figure 4.1 the next system(s) is defined as ASC, asc, C, and is partly hatched to indicate that the first of these systems is sodium dependent. In fact these three systems show great homology in their substrate specificity, but great differences in affinities and cation-dependence. In human cells, the Na^+ dependent ASC system is the principal route for cysteine entry at physiological plasma levels of cysteine (the apparent K_m being 15 μM). There is a significant affinity for alanine and serine (hence the acronym), but not for glycine or for larger neutral amino acids. The system can operate in both a net and an exchange mode, and Al-Saleh & Wheeler (1982) emphasised that exchange is predominant in human red cells. A coupling ratio of 1:1 has been demonstrated for Na:alanine, and surprisingly the System does not appear to be electrogenic, in that it is unaffected by marked changes in the membrane potential (Ellory *et al.*, 1980 – see Lerner, 1985). Ellory *et al.* (1982) showed that System ASC but not System L was sensitive to inhibition by NEM and PCMBS, in contrast to the effects of loop diuretics (particularly MK-196) which are more effective inhibitors of System L than System ASC.

4.4.3 *Neutral amino acids: System C*

Aside from the ubiquitous capnophorin, System C is the only amino acid transport system present in sheep red cells. Further, mutants exist which lack the C transport system and contain low levels of GSH and high concentrations of basic amino acids. In fact the lesion was first identified by the analysis of red cell Na^+ and K^+ levels (Ellory *et al.*, 1972 – see Ellory & Young, 1982) which indicated the need for another major intracellular cation (up to

25 mM ornithine and lysine). The sheep C System shows the same pattern of substrate specificities as the ASC System, but with much lower affinities (thus the K_m for cysteine and alanine on System C is 12 and 16 mM, respectively, whereas for these amino acids on System ASC they are 15 and 150 μM). System C has a significant but low affinity for ornithine and lysine, but not for arginine. This accounts for the presence of these dibasic amino acids intracellularly as a relic of reticulocyte maturation in animals lacking the transporter. It also fits in with the idea that C may be a modified ASC System without its Na^+ dependence, but with the amino group of the basic amino acids substituting for Na^+ at a binding site in some cases. A recent exciting finding which is represented in Figure 4.1 as asc, is that horse red cells also possess Na^+ independent amino acid transport systems with the same substrate specificities as ASC and C (Fincham *et al.*, 1985). Here the situation is more complex than in the sheep erythrocyte, since although, as in the case of sheep, the inheritance data indicate that the trait is genetically determined, the system is triallelic, and therefore two transporters defined as asc-1, and asc-2 have been proposed. The substrate specificities of these two systems again coincide, but it is the affinities which are different, with asc-1 corresponding to the relatively high affinities found for ASC (but remaining Na^+ insensitive) whereas asc-2 shows low affinities more equivalent to the sheep C System. How these systems are related in molecular terms remains an intriguing puzzle, the resolution of which will require the use of molecular genetics.

4.4.4 *Neutral amino acids: System L*

The L System is the red cell amino acid transporter with the highest transport capacity, the widest species (and tissue) distribution, and for which there exist the best kinetic data (see section 4.3 above). As Figure 4.1 clearly shows, its preference is for medium-sized neutral amino acids. Interestingly the system is not stereospecific, showing a 3.5-fold lower affinity for D-leucine *vs* L, but a twofold greater V_{max} (0.21 *vs* 0.12 mol $\cdot$ l^{-1} cells $\cdot$ h). These V_{max} figures also emphasise the high capacity of this system, and indicate that even a poor substrate may achieve a significant flux via System L because of the relative abundance of the transporter in the cell membrane. This is shown in Figure 4.2, where three non-ideal L System substrates all have significant fluxes by this system. L System operates in either exchange or zero-trans modes, and represents the major route for amino acid uptake in most species' red cells. Exceptions noted to date include ruminants, horses and cats.

4.4.5 *Neutral amino acids: Systems T and N*

Following the discussion of Systems L and ASC, which show a broad substrate specificity (Figure 4.2) we come to four specialised systems. System T, the first of these shows a significant transport affinity for only the aromatic amino acids tryptophan, tyrosine and phenylalanine. In fact, even for tryptophan, its ideal substrate (at least acronymically) this system only accounts for 70% of the flux at plasma concentrations around 0.1 mM (Figure 4.2). Although the evidence for this system is based purely on kinetic experiments, it has been confirmed in three separate studies, and the existence of this rather restricted System T as a separate entity, which seems to be confined so far only to human red cells seems secure.

The N System has been defined previously by Kilberg *et al.* (1980) in hepatocytes, and its presence in human red cells confirmed by kinetic studies using glutamine as the principal substrate (Ellory & Osotemehin, 1983). Our original idea that this system is the source of glutamate for GSH biosynthesis, because of the absence of glutamate transport in human red cells, has been disproved by our own unpublished incorporation studies using labelled glutamine, and NMR studies by King & Kuchel (1985) which indicate there is no intracellular glutaminase activity. Thus, although this system is Na^+ dependent, and has relatively high affinities for its substrates (glutamine, asparagine and histidine), no obvious role can be assigned to it at present in the mature red cell.

4.4.6 *Cationic amino acids*

The y+ System (formerly called ly+) is a selective, high affinity system which has been identified in human and carnivore red cells (dog, cat) with a specificity for the basic amino acids lysine, ornithine and arginine. The system is unusual in having high substrate affinities (apparent K_m for arginine 50 μM) but it is not

Na^+ dependent. This may be a complementary role for the terminal amino group at a Na^+ binding site as suggested earlier in connection with the effects of these basic amino acids on the ASC-type Systems.

4.4.7 *Anionic amino acids*

Although the red cells of most species are impermeable to acidic amino acids, carnivore red cells possess a Na^+ dependent high affinity transport system for aspartate and glutamate (Ellory *et al.*, 1982). This has recently become of considerable interest with the discovery by Maede *et al.* (1982) of a hereditary trait in dogs which increases the transport activity of this system enormously, leading to high intracellular levels of glutamate, aspartate and GSH, and changes in the high Na^+/low K^+ status of these red cells. The kinetic characteristics of the glu system (called X^- in Table 1.1) are very similar to its activity in nervous tissue, leading to the proposal that carnivore red cells may be an ideal model for studying this important transport system.

4.5 **Transport of oligopeptides**

This has now completed the survey of known amino acid transport systems presented in Figure 4.1. There are however, two recent papers which may be of great significance in understanding amino acid transport in red cells. Kuchel's group (King & Kuchel, 1984; 1985), using NMR, have demonstrated significant uptake and hydrolysis of dipeptides by red cells. This immediately suggests glutamyl peptides as the source of glutamate for GSH* biosynthesis, and provides a dramatic new digestive role for erythrocytes, in terms of scavenging peptides and hydrolysing them. The amino acid transport systems could then conceivably even be acting in reverse, i.e. exporting amino acids derived from peptide hydrolysis from the cell. This is clearly a major new role for red cells and much work needs to be done to confirm and establish the principles of peptide uptake by erythrocytes. Some of the possibilities have far-reaching consequences, particularly when one considers the number of small peptide hormones potentially present in the blood. This is certainly the most provocative suggestion since Elwyn's proposals on interorgan transport (Elwyn 1966; see also Christensen, 1982).*

4.6 Biochemistry of amino acid carrier systems in the red cell

Aside from these ideas, what conclusions can we draw from studying red cell amino acid transport? As a model, red cells can provide high quality kinetic data, such as Rosenberg's L System results which help both modelling of carriers, and predicting the behaviour of systems like L in other cell types. In terms of red cell biochemistry, the origin of at least two of the three substrates for GSH biosynthesis (glycine, cysteine) has been determined in detail – that of glutamate remains a fascinating matter of debate. As dead-end cells, which do not have to perform protein synthesis, red cells can tolerate potentially lethal disturbances of amino acid transport compared with more conventional cells. This is reflected in the three Systems already described in the literature with genetic variations (sheep, horse, dog). These intraspecific variations offer certain biochemical and particularly immunological possibilities for studying transporters. This brings us back to the overwhelming problem in studying amino acid transport systems: the lack of specific inhibitors. A few biochemical studies have been initiated, on transporter solubilisation and characterisation. Radiation inactivation analysis of systems L and asc indicates apparent molecular weights of 120,000–150,000 Da suggesting a possible correspondence with a dimer of Band 4.5 as for the nucleoside and glucose transporters. There is however a long way to go before we can describe either the structure or the functional mechanism of any of the amino acid transport systems presented in this short review.

---

*In placental brush border membrane vesicles the poorly hydrolysed dipeptide glycyl-L-proline inhibited proline uptake (Boyd & Lund, 1981; see Chapter 5.4). For review of transmembrane transport of small peptides and its relation to amino acid transport see Mathews & Payne, 1980. Transport of the tripeptide glutathione which occurs in most living organisms, has been recently reviewed (Bannai & Tateichi, 1986). (Editor's note.)

4.7 **Further reading**

(Other literature is listed in the compiled References at the end of the book.)

Christensen, H. N. (1982). Interorgan amino acid nutrition, *Physiological Reviews*, **62**, 1193–1233.

Ellory, J. C. & Lew, V. L. (1977). *Membrane transport in red cells.* Academic Press, New York.

Ellory, J. C., Jones, S. E. M. & Young, J. D. (1981). Glycine transport in human erythrocytes. *Journal of Physiology*, **320**, 403–22.

Ellory, J. C. & Young, J. D. (eds.) (1982). *Red cell membranes – a methodological approach.* Academic Press, London.

Fincham, D. A., Mason, D. K., Paterson, J. Y. F. & Young, J. D. (1987). Heterogeneity of amino acid transport in horse erythrocytes. A detailed kinetic analysis of inherited transport variation. *Journal of Physiology* (London) (in press).

5

Blood-tissue movement of amino acids in various organs: the basolateral membrane and its importance in the polarity of transepithelial transport

C. A. R. Boyd Department of Human Anatomy Oxford University *and* **D. L. Yudilevich** Department of Physiology King's College University of London

5.1 **Serial barriers to blood–tissue transport**

In considering the movement of substrates into or out of a tissue it is important to think of the various and successive membranes and barriers which are present in that particular organ (as discussed in Yudilevich & Mann 1985). A simplified scheme would take into account serial barriers such as the capillary endothelium with its luminal (blood facing) and antiluminal (interstitial) sides, and the interstitial facing membrane of the parenchymal cell, through which molecules are transported into the cytoplasmic compartment. Non-parenchymal cells such as glia, fibroblast and Kuppfer cells, form 'minority' groups which are usually either ignored or have to be studied following their isolation.

The amino acid composition of the plasma and that of the intracellular fluid, in two different tissues (placenta and heart), are given in Table 5.1 where the ability of cells to accumulate amino acids, albeit to varying extents, is clearly illustrated (e.g. glu, asp, gln, tau). The aggregate concentration of amino acids belonging to a group, presumably transported by a given transporter (see Table 1.1), may be obtained from inspection of Table 5.1. These intracellular concentrations should be considered since changes in individual amino acids will influence the influx of other amino acids into the tissue from the plasma.

Table 5.1 *Amino acid concentrations in human maternal plasma fetal/maternal plasma concentration ratios and placental tissue concentrations*[a]

Amino acid	Human maternal plasma [rat normal plasma] mmol · l^{-1}	Fetal/maternal plasma	Placenta tissue human [rat heart][d] (mmol · kg^{-1} wet wt)
Leucine, Leu	0.06 [0.12]	1.98	0.38 [0.14]
Isoleucine, Ile	0.03 [0.07]	2.00	0.17 [0.09]
Valine, Val	0.10 [0.16]	2.04	0.30 [0.14]
Methionine, Met	0.02 [0.04]	1.28	0.06 [0.10]
Tryptophan, Trp			
Tyrosine, Tyr	0.03 [0.06]	2.03	0.17 [0.10]
Phenylalanine, Phe	0.04 [0.05]	2.00	0.19 [0.85]
Glutamine, Gln[b]	0.32 [0.94]	1.37	2.23 [8.70]
Asparagine, Asn	0.16 [—]	1.33	— [—]
Histidine, His	0.07 [0.08]	3.95	0.17 [0.35]
Alanine, Ala	0.26 [0.31]	1.61	0.97 [1.90]
Serine, Ser	0.10 [0.16]	1.43	0.57 [1.10]
Cysteine, Cys	— [0.10]	—	— [0.09]
Threonine, Thr	0.06 [—]	1.63	0.54 [—]
Glycine, Gly	0.14 [0.20]	1.63	1.34 [0.89]
Proline, Pro	— [0.14]	—	— [2.40]
Aspartate, Asp	0.02 [0.01]	1.05	1.05 [2.10]
Glutamate, Glu[b]	0.10 [0.16]	1.31	0.69 [7.60]
Lysine, Lys	0.10 [0.35]	2.95	0.62 [0.50]
Arginine, Arg[c]	0.04 [0.11]	2.12	0.28 [0.09]
Ornithine, Orn	0.05 [0.05]	1.98	— [0.17]
Citrulline, Cit	0.01 [—]	0.93	— [—]
Taurine, Tau[b]	0.06 [0.05]	2.25	— [29.0]

[a](i) Placental data, from references in Table 1 of Yudilevich & Sweiry (1985).
(ii) Data for the rat (Baños *et al.*, 1978) is added here to illustrate characteristic concentrations in tissues other than placenta.
(iii) For data on intracellular amino acid concentrations in several tissues see Holden (1962). Amino acid concentrations in twelve tissues of the normal and fasting rat were measured by Herbert *et al.* (1966).
(iv) The grouping in which these amino acids are listed relates to specific transport systems (see Table 1.1). This hints at the concept that the added concentrations are of importance in respect of movement of a single amino acid by its carrier.

[b]Glutamine, glutamate and taurine concentrations are very high in most tissues, e.g. glutamate in brain is the highest (9 mmol · l^{-1}).

[c]Arginine is barely detectable in rat liver (0.007 mmol · g^{-1} wet wt) (White, 1985).

[d]As mmol · l^{-1} intracellular water.

*For placenta, combined asn-gln; for heart, gln alone.

5.2 Experimental approaches to amino acid transport in organs

In vivo experiments in perfused intact organs provide conditions appropriate for blood–tissue transport studies (see Yudilevich & Mann, 1985) but biochemical techniques such as the preparation of membrane vesicles enriched with plasma membranes from the basolateral side of the cell allow more precise characterisation for example of the driving forces for a given transport process. There are, however, pitfalls in the use of vesicles which include the question of the orientation (inside or outside out) of the vesicle, variation in vesicle size, vesicle 'leakiness', and the absence of the normal cytoplasmic metabolic machinery.

The value of K_m is useful in characterising transport under the conditions in which it has been studied. Table 5.2 shows a selection of kinetic measurements for transport at the 'blood' side of cells in various organs. The 'mutability' of these determinations was pointed out many years ago by Fisher (1964). This point needs particular emphasis today, when the variety of different experimental techniques has broadened dramatically. Therefore in making use of K_m and V_{max} to investigate regulatory mechanisms or the action of various effectors the importance of appropriate and adequate controls needs special emphasis. For example in considering the extent to which any experimentally observed K_m for any given amino acid will be an over-estimate of the real half-saturation constant of the membrane transport system under study it is important to realise that the correction factor (Vh/D, where $V = V_{max}$, h the thickness of the unstirred layer, and D the diffusion coefficient of the amino acid under study) will be a linear function of the V_{max} of the system. Thus it is readily apparent that alterations which are observed empirically in K_m may be a consequence of changes in h (for example because of changes in tissue structure or blood flow) as well as of changes in V_{max} and need not necessarily be the result of any change at all in the affinity of the membrane transporter. This problem will be more marked for systems which are less well stirred, for example in experiments on tissue slices.

5.3 Epithelia may transport amino acids mainly for their own metabolism or to produce net transcellular (polar) transfer

With respect to tissue transport of amino acids in animal cells, tissues may be separated into those designed around polar epithelial cells (for example those of the kidney tubule, intestinal epithelium, liver, exocrine glands and skin) and those with parenchymal cells which are not polarised, such as muscle and nervous tissue. In the latter, the parenchymal cells do not effect net vectorial transcellular transport. The former group may be further subdivided between epithelia in which much of the transport of amino acid is vectorial, producing net transcellular movement (kidney, intestine, placenta) and the group in which transport from the interstitium into the cell is important but the plasma membrane on the opposite side is essentially impermeable (for example liver, exocrine organs and skin).

The transport of blood-borne substrates of a lipid-insoluble nature and of relatively small size such as the amino acids is generally not limited by the capillary barrier since in general these solutes diffuse across the inter-endothelial space quite freely (Yudilevich, 1985). Specific mechanisms for transport are therefore located in the basolateral side (interstitium-facing side) of the parenchymal cells and such mechanisms as do exist in the capillary endothelium may be hard to detect. An exception (which is discussed below) occurs in the brain capillary in which the endothelium is particularly and unusually tight (blood-brain barrier) and the endothelial cell membrane is the site at which specific transport mechanisms facilitate the movement of amino acids from the blood into the brain interstitium.

5.4 Amino acid transport in the small intestine

The brush border surface of the small intestinal epithelium comes in contact with α-amino-nitrogen from two sources: that ingested in the food, and that found in the endogenous secretions (including desquamated cells) of the upper gastrointestinal tract. Such α-amino-nitrogen may be in the form of large proteins, peptides (of larger or smaller size) and some free amino acid residues. Fisher (1954) was one of the first to re-emphasise the importance of peptide rather than amino acid absorption across this surface since the rate of protein digestion within the lumen of the gastrointestinal tract is too slow to allow complete hydrolysis of all peptide bonds in ingested protein. It is now clear that oligopeptides are the major form in which amino-nitrogen enters the small intestinal epithelium and it has recently been shown that the electrogenic transport of such peptides (Boyd & Ward, 1982) is coupled to the electrochemical gradient of H^+, rather than to that of Na^+, across the brush border membrane (Hoshi, 1985; see Chapter 2).

However, of interest to this particular chapter is the movement of the amino acids across the basolateral surface of this epithelium. Net flux of amino acids will of course be outward (from epithelium to blood) during periods of absorption but during post-absorptive periods can be in the opposite direction (from blood to epithelium). Surprisingly, until recently rather little was known of the transport events occurring at this particular plasma membrane which Freel & Goldner (1981) described as 'the dark side of the epithelium'. Two technical developments have recently allowed some features of these transport processes to be examined.

The application of membrane vesicles to the small intestine (Murer & Hopfer, 1976) was a classic example of the value of biochemistry in understanding a physiological process. Shown in Figure 5.1 is the transport of valine into both brush-border and basolateral membrane vesicles. In the presence of an inwardly directed sodium gradient the brush border membranes show clear evidence of a transient accumulation of the amino acid ('the overshoot') as sodium ions enter the vesicle with valine through a co-transport system – precisely which system is involved is not yet clear. This transient accumulation is not maintained since once the sodium electrochemical gradient is dissipated in such an isolated membrane system and in the absence of ATP, there is no possibility of establishing the dynamic steady state seen *in vivo* and, as a result, the intravesicular valine concentration then gradually equilibrates with that in the external medium. However when the identical experiment is performed on vesicles prepared from the basal surface of the epithelial cells (marker enzymes being used to define precisely the source of each membrane fraction) no such transient overshoot of intravesicular accumu-

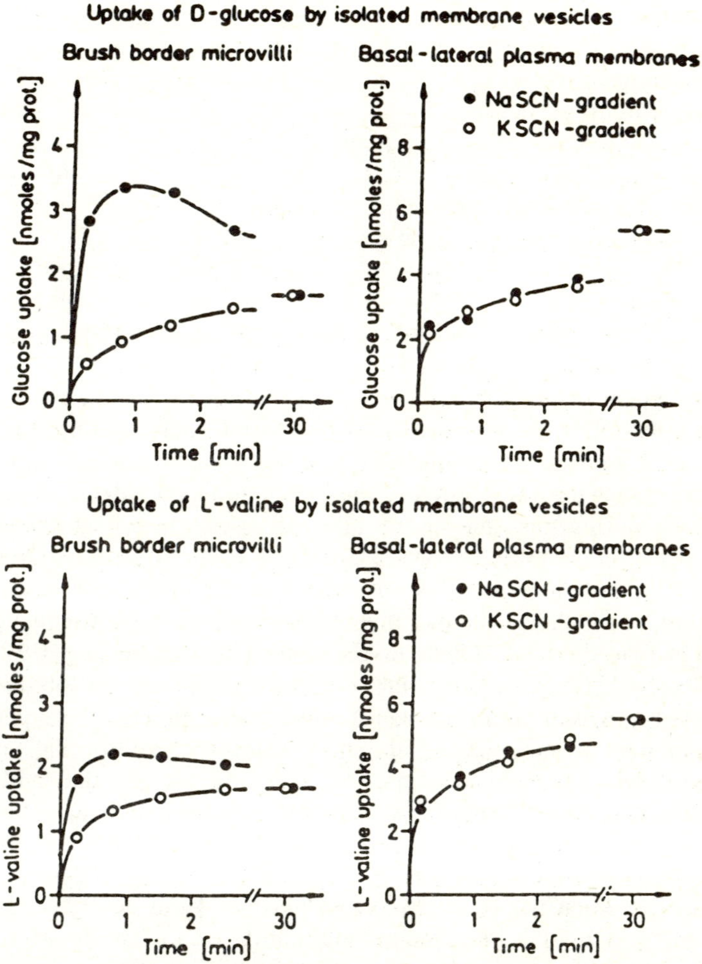

Fig. 5.1 Asymmetry of sodium dependence of valine transport in small intestinal brush-border and basolateral membrane vesicles. Na^+ dependence and independence of valine transport in membrane vesicles isolated respectively from the brush border and basolateral surfaces of the rat small intestinal epithelium. Also shown are results for glucose transport. (Taken with permission from Hopfer, U. *et al.* (1976). *Journal of Cellular Physiology*, **89**, 805–10.)

lation of valine is seen even in the presence of an inwardly directed sodium gradient. Polarity of net trans-epithelial valine movement from lumen to blood may therefore be seen to be a consequence of the asymmetric distribution of a 'valine pump' (a secondary active transporter coupled to the sodium electrochemical gradient) and of a 'valine leak' (a facilitated diffusion system for the amino acid, probably System L) between the two faces of the cell.

The other type of experiment which has elucidated some of the features of the basolateral membrane is that of vascular perfusion. Shown in Figure 5.2 is the result of an experiment in which the non-metabolised amino acid aminoiso butyric acid (AIB) has been studied. AIB initially was added to the lumen of the intestine, subsequently abruptly removed from the lumen and the ensuing washout of the amino acid into the portal venous effluent was studied as a function of time. During this washout another amino acid, L-leucine, was added to the lumen and as the figure clearly shows during this time the rate of AIB appearance in the blood decreases markedly. D-leucine failed to give such an effect. Other experiments show that this was the result of an inhibition of AIB efflux across the basolateral membrane rather than a consequence of increased AIB backflux into the lumen (Boyd & Perring, 1982). Clearly this basolateral membrane has transporters (in this case possibly System A or L) which show rather precisely the phenomenon of competitive inhibition which is stereospecific and dependent upon the concentration of the inhibitory amino acid.

In addition to showing inhibitory interactions, amino acid efflux across the basolateral membrane may be stimulated by the addition of other amino acids. Such an experiment is shown in Figure 5.3 in which lysine efflux (the rate here being described by a rate constant) is stimulated by addition of leucine to the arterial infusate of the vascularly perfused intestine. This effect again was a function of the concentration of the elicitor and was also stereospecific (Cheeseman, 1983). The unexpected effect between different classes (basic and neutral) of amino acid may be classified as hetero-cis inhibition and hetero-trans stimulation. They must be important in the processes by which absorbed amino-nitrogen (following its final intra-epithelial digestion to produce free amino acids from the oligopeptide mixture taken up across the brush border membrane) leaves the intestinal epithelial cell. Much still

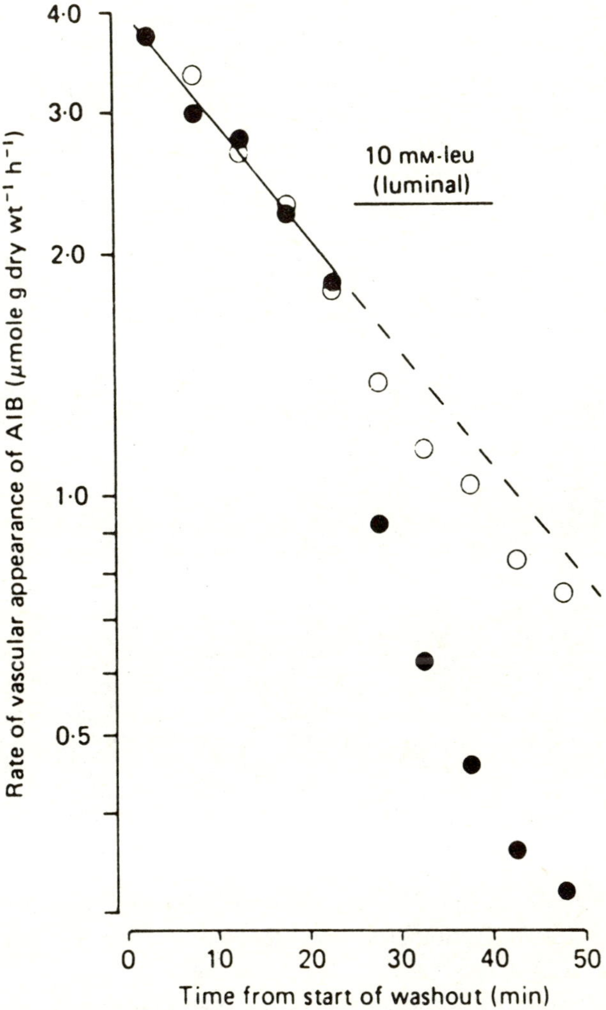

Fig. 5.2 L-leucine inhibition of aminoisobutyric acid (AIB) transport across the basolateral membrane of the vascularly perfused small intestine. Influence of leucine (leu; 10 mM) added to the intestinal lumen upon the washout of AIB into the vascular bed from the pre-loaded small intestinal epithelium. Two washouts were performed following identical loading with AIB and the results are superimposed. During the first (•) L-leucine and during the second (○) D-leucine was added to the lumen for the period indicated. (Taken from Boyd C. A. R. & Perring, V. S. (1982). *Journal of Physiology*, **327**, 53–64.)

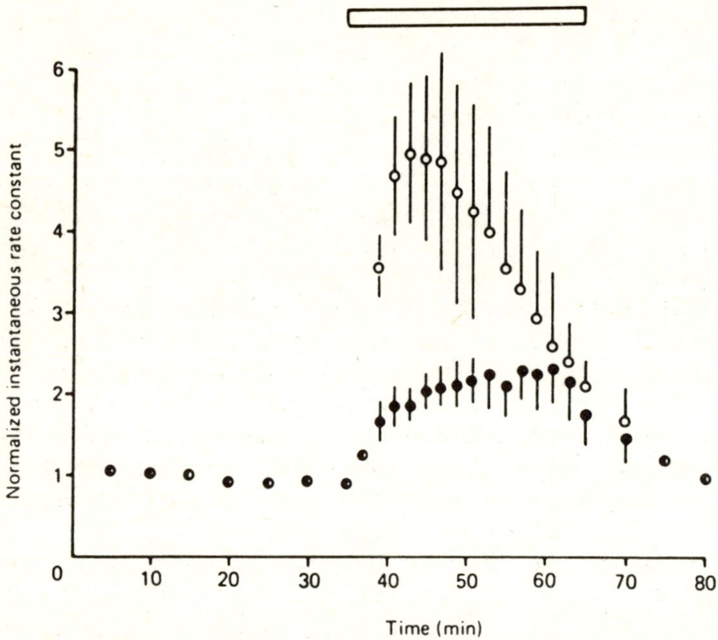

Fig. 5.3 Effect of L-leucine on efflux of lysine across the basolateral membrane of the perfused small intestine. The rate constants for the washout of lysine from the small intestinal epithelium is shown, each point representing the mean of four experiments (vertical bars show the standard error of the mean). The horizontal bar indicates leucine addition (0.1 mM (○); 0.01 mM (•) to the vascular bed. A similar stimulation of lysine washout was seen when leucine was added to the lumen. (Taken from Cheeseman, C. I. (1983). *Journal of Physiology*, **338**, 87–97.)

remains to be learnt both about the transport systems involved and the underlying physiological importance of these interactions.

With regard to transport across the whole epithelium rather than across individual membranes, the presence of sodium ions in the intestinal lumen are of importance for the sodium-coupled entry of amino acids at the brush border membrane but probably are not of direct energetic importance in the movement of amino acids across the basolateral membrane, although sodium pumping may be required to keep the spaces between adjacent epithelial cells open. However Table 5.2 shows that sodium dependence is a feature found at the basolateral side of epithelia almost as frequently as at the brush border.

Shown in Figure 5.4 are the results of an experiment on the transport across the epithelium of various concentrations of leucine with and without sodium present in the lumen. The transport of leucine saturates as a function of luminal amino acid concentration and removal of sodium ions from the lumen causes a marked reduction in the maximal rate of transfer (V_{max}) whilst having a smaller effect on the apparent affinity (the K_m rising from 2.1 to 4.5 mM upon removal of sodium). Other experiments showed that removal of sodium from the lumen caused a greater inhibition (43%) than the removal of sodium from the vascular bed (27% inhibition). Complete removal of sodium from both sides of the intestinal epithelium reduced leucine transfer to 23% of that in the presence of sodium. This suggests that System L performs a more important role in transepithelial amino acid transport at the basal than it does at the apical surface of the epithelium, in keeping with recent results on membrane vesicles (Mircheff *et al.*, 1980; see Stevens *et al.*, 1984).

In summary, our present knowledge suggests that multiple systems are responsible for amino acid transport in both faces of the epithelium (see Munck, 1981). Stevens & Wright (1985) describe at least seven systems in the brush border membrane with overlapping specificities. For neutrals three Na^+ co-transporters are proposed (Wright *et al.*, in Kilberg, 1986). Furthermore the role of K^+ in histidine and glutamine transport in the brush border has been emphasised by Giordana *et al.*, 1985. The transporters in the basolateral membrane which have been described in studies on vesicles, are System 1 (equivalent to ASC), System 2 (equivalent to L), System 3 (reactive with alanine and serine but not with glycine or threonine), and System 4 (gly and thr but not ala or ser) (Lash & Jones, 1984). They are not all Na^+ independent (see Table 5.2). Much remains to be done in this field, in particular with regard to how the epithelium behaves when presented with physiological concentrations of amino acids at the mucosal and serosal faces simultaneously. The description of an inherited abnormality in children of lysine transport across the basolateral membrane (Rajantle *et al.*, 1980) points to the opportunity which study of the molecular genetics of such systems offers. Table 5.3 show other genetic disorders affecting amino acid transport (see Wellner & Meister, 1981; Scriver & Tenenhouse, 1985; Benson & Fenson, 1985).

Table 5.2 *Kinetic constants and sodium dependence of amino acid transport at the blood – tissue interface in different organs*

Amino acid	Experimental method	K_m ($mmol \cdot l^{-1}$)	V_{max}
Intestine (basolateral side)			($nmol \cdot min^{-1} \cdot mg^{-1}$ protein)
Alanine	Rat vesicles (Mircheff *et al.*, 1980)[a]	0.73 (+)	5.3
Cysteine	Rat vesicles (Lash & Jones, 1984)	0.35 (−)	2.0
Valine	Rat vesicles (Hopfer *et al.*, (1976)	5.80 (−)	43.0
Kidney (basolateral side)			($nmol \cdot h^{-1} \cdot mg^{-1}$ DNA)
AIB	Rat cells (Rabbito & Karsh, 1982)	4.47 (+)	3.9
			($nmol \cdot min^{-1} \cdot 10^{-6}$ cells)
Alanine	Pig cells (Sepulveda & Pearson, 1982)[b]	0.28 (+)	7.1
Leucine	Pig cells (Sepulveda & Pearson, 1982)[b]	0.06 (−)	4.6
Glutamate	Rat vesicles (Sacktor *et al.*, 1981)	(+)	
Alanine	Rat vesicles (Murer & Kinne, 1975)[c]	8.60	
Lysine	Pig cells (Sepulveda & Pearson, 1985)	0.30 (−)	0.6
Placenta (maternal and fetal sides)			($\mu mol \cdot min^{-1} \cdot g^{-1}$)
Alanine	Perfused, maternal side in guinea-pig (Eaton *et al.*, 1982; Wheeler, 1986)[d]	10.30 (+)	14.5
Alanine	Perfused, fetal side in guinea-pig (Eaton *et al.*, 1982; Wheeler, 1986)[d]	8.40 (−)	4.4
Phenylalanine	Perfused, maternal side in guinea-pig 3.10 (Eaton *et al.*, 1982)	3.7	
Phenylalanine	Perfused, fetal side in guinea-pig (Eaton *et al.*, 1982)	11.90	9.5

Lysine	Perfused, maternal side in guinea-pig (Wheeler & Yudilevich, 1984)[e]	1.70	1.9
Lysine	Perfused, fetal side in guinea-pig (Wheeler & Yudilevich, 1984)[e]	0.90 (−)	
		($nmol \cdot l^{-1}$ IVW $\cdot h^{-1}$)	
Alanine	Human brush border vesicles (Assai *et al.*, 1982)[e]	0.38 and 3.50 (+)	0.18 and 1.83
AIB	Human brush border vesicles (Ruzycki *et al.*, 1978)	1.00 (+)	12.5
		($mmol \cdot min^{-1} \cdot mg^{-1}$ protein)	
Proline	Human brush border vesicles (Boyd & Lund, 1981)	1.00 (+)	0.7–20
Salivary gland (basolateral side)			
Alanine	Perfused gland in cat (Mann *et al.*, 1984)	0.83 (+)[a]	0.06
Serine	Perfused gland in cat (Mann *et al.*, 1984)	1.20 (+)	1.00
Phenylalanine	Perfused gland in cat (Mann *et al.*, 1984)	6.40 (−)	1.72
Lysine	Perfused gland in cat (Mann *et al.*, 1984)	0.49 (−)	0.23
AIB	Perfused gland in cat (Keryer & Rossignol, 1980)	4.89 (+)	0.73
AIB	Rat vesicles (Takuma & Baum, 1985)	1.30 (+)	0.78
Pancreas (basolateral side)		($\mu mol \cdot min^{-1} \cdot g^{-1}$)	
Serine	Perfused gland in rat (Mann & Peran, 1986)	14.40 (−)	25.90
MeAIB	Perfused gland in rat (Mann & Peran, 1986)	1.70 (+)	0.49
Alanine	Cat vesicles (Tyrakowski *et al.*, 1978)[f]	(+)	
Alanine	Mouse acinar cells	1.60 (+)	

Table 5.2 (*continued*)

Amino acid	Experimental method	K_m ($mmol \cdot l^{-1}$)	V_{max}
Liver		($nmol \cdot min^{-1} \cdot mg^{-1}$ protein)	
AIB	Rat hepatocyte (LeCam & Freychet, 1977)	25.00	6.5
AIB	Rat hepatocyte (Bracy *et al.*, 1986)	0.89 (+)	4.0
Taurine	Rat heaptocyte (Hardison & Weiner, 1980)	0.04 (+)	0.3
BCH	Rat hepatoma (Nadziejko & Reichberg, 1980)	0.13 (−)	0.3
Alanine	Rat sinusoidal side vesicles (Van Amelsvoort *et al.*)	3.00 (+)	
		($\mu mol \cdot min^{-1} \cdot g^{-1}$ICW	
Glutamine	Rat isolated hepatocytes (Kilberg *et al.*, 1980)[g]	1.10 (+)	8.0
		($pmol \cdot min^{-1} \cdot mg^{-1}$ protein)	
Glutamate	Rat vesicles (Sips *et al.*, 1982)	0.0074 (+)	0.33
		($nmol \cdot min^{-1} \cdot mg^{-1}$ protein)	
Lysine	Rat hepatoma (White, 1985)	0.13 (−)	3.26
Blood–brain barrier (luminal side)			($nmol \cdot min^{-1} \cdot g^{-1}$)
Phenylalanine	Rat *in vivo* (Pardridge, 1983)	0.11	28.0
Leucine	Rat *in vivo* (Pardridge, 1983)	0.10	22.0
Tryptophan	Rat *in vivo* (Pardridge, 1983)	0.16	30.0
Arginine	Rat *in vivo* (Pardridge, 1983)	0.09	7.8
BCH	Rat *in vivo* (Wade & Katzman)[h]	0.16	
			($\mu mol \cdot min^{-1} \cdot S^{-1}$)
Leucine	Dog, perfused (Betz *et al.*, 1975)[i]	1.58	0.32
BCH	Dog, perfused (Betz *et al.*, 1975)[i]	0.16	

Blood–brain barrier (anti-luminal side)		(nmol · min^{-1} · mg^{-1} protein)	
Phenylalanine	Rat isolated caplillaries (Cardelli *et al.*, 1981)[i]	0.12	0.09
Lysine	Rat isolated capillaries (Cardelli *et al.*, 1981)[i]	0.10	0.11
Cystine	Rat isolated capillaries (Hwang *et al.*, 1980)	0.04	
Leucine	Rat isolated capillaries (Hjelle *et al.*, 1978)[i]	0.13 (−)	

[a]See Stevens *et al.*, 1984; [b] see Sepulveda & Pearson, 1985; [c]see Shafter & Williams,; [d]see Wheeler, C. P. D., 1986; [e]see Yudilevich & Sweiry, 1985; [f]see Takuma & Baum, 1985; [g]see Kilberg, 1985; [h]see Christensen, 1979; [i]see Pardridge, 1983.
AIB = aminoisobutyric acid; MeAIB = methyl AIB; BCH = 2-amino-norborane-2-carboxylic acid; IVW = intra vesicular water; ICW = intra cellular water.

(+) or (−) refers to the presence or absence of sodium dependence.

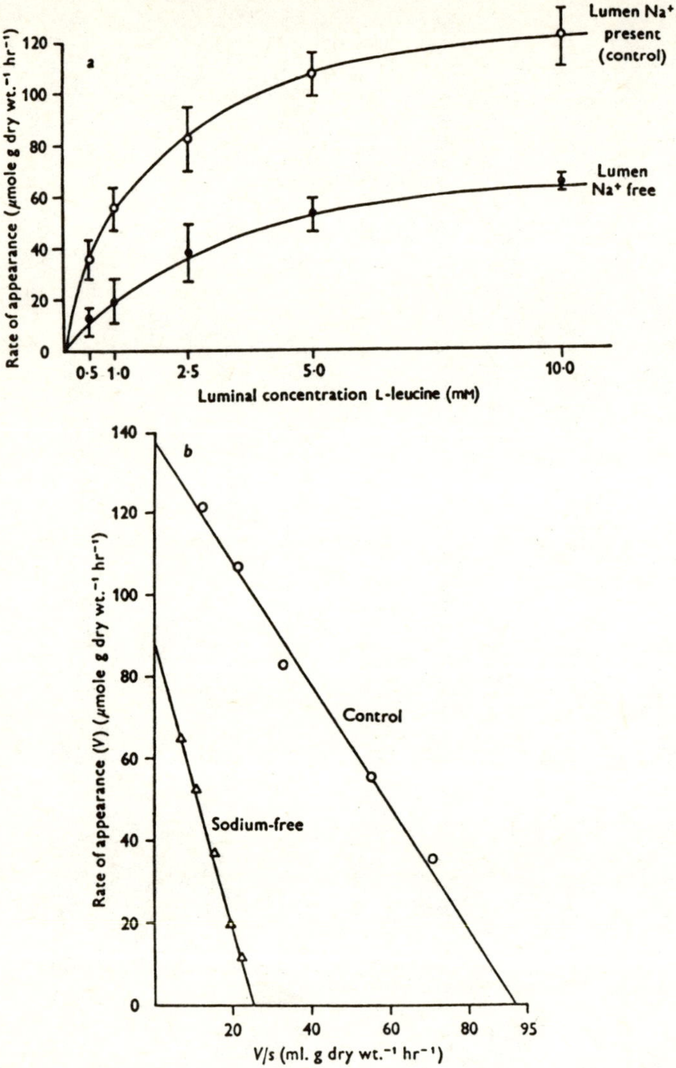

Fig. 5.4(*a*) Influence of sodium added to the lumen on transport of leucine across the small intestinal epithelium (from lumen to blood) in the perfused small intestine. Control curve (○) from five experiments, sodium-free (K^+ substitution) curve (●) derived from three experiments. Values are means ± SEM.

(*b*) Hanes plot of the data presented in Figure 5.4*a*. *S*, concentration of L-leucine in lumen (mM), *V*, rate of leucine appearance in blood: (○) control data; (△), sodium-free data. Note the 'mixed' inhibitory effect of Na removal with both a reduction of V_{max} and an increase in 'K_t' (Taken from Boyd, C. A. R. *et al.* (1975). *Journal of Physiology*, **250**, 409–29.)

Table 5.3 *Amino acid transport systems of human kidney revealed by mutation*[a]

Substrates for the transport system	McKusick No.		
Cys, lys, orn, arg	22010	Cystinuria	BBM; mutation also expressed in small intestine
Cys	23820	Isolated cystinuria	BBM
Lys, orn, arg	22270	Dibasicaminoaciduria I (Lysine-protein intolerance)	BLM; mutation also expressed in small intestine (and liver?)
Lys, orn, arg	12600	Dibasicaminoaciduria II	BBM?
Asp, glu	23615	Dicarboxylicaminoacid uria	Mutation also expressed in small intestine
Pro, hypro, gly	24260	Iminoglycinuria	Mutation also expressed in small intestine
Neutral amino acids (exclusive of above groups)	23460	Hartnup disorder	BBM; mutation also expressed in small intestine

McKusick numbers refer to listings in Catalogues of Mendelian Inheritance in Man (McKusick, V. A. (ed.) 6th Edn Johns Hopkins Press, Baltimore, 1983). Autosomal dominant inheritance, 10 000 series; autosomal recessive, 20 000 series; X-linked, 30 000 series.

BBM = brush border membrane; BLM = basolateral membrane

[a]Modified from Scriver C. R. & Temenhouse H. S. (1985). *Annals of the New York Academy of Sciences*, **456**, 384–397.

5.5 Amino acid transport in the kidney

There are many structural and functional features in common between the epithelium of the small intestine and that of the renal proximal tubule. This is clearly shown by the finding of identical abnormalities of amino acid transport in the brush border membrane of the two tissues in specific disorders of amino acid transport which have a genetic basis (e.g. Hartnup disease and cystinuria, see Table 5.3). The basolateral membrane of the renal proximal tubule has also been studied in this tissue with isolated membrane vesicles (Table 5.2) and also electrophysiologically (Samarzija & Fromter, 1982). Figure 5.5 shows the

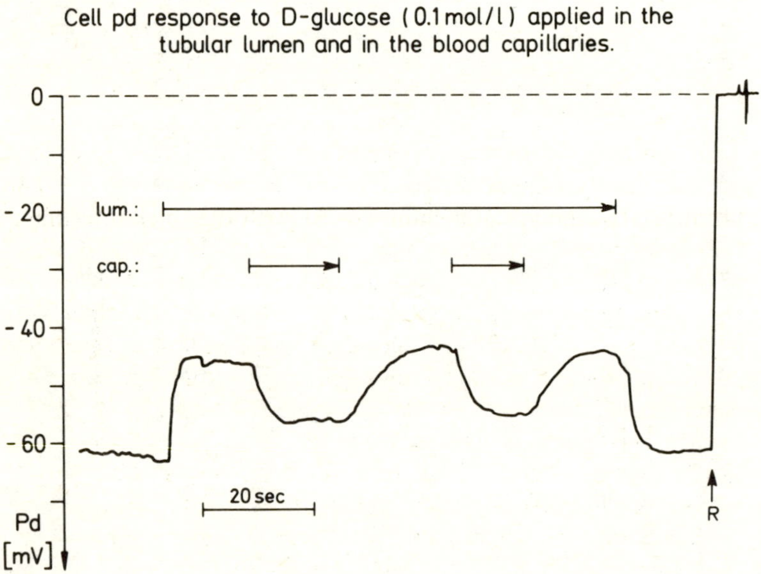

Fig. 5.5 Effect of peritubular (basolateral) D-glucose perfusion during simultaneous luminal D-glucose perfusion, on proximal tubular cell membrane potential. For a period (shown by the long arrow) the tubular lumen was perfused with a solution containing 100 mM D-glucose, which partially depolarised the cell. During the times indicated by the short arrows the peritubular (basolateral) capillary perfusion was switched from a control solution (containing an appropriate osmotic substitute) to one containing 100 mM D-glucose. It can be seen that during this time the cell gradually repolarises; after cessation of peritubular D-glucose perfusion the potential returns to its original depolarised state. (Taken from Samarzija, J. & Fromter, E. (1982). *Pflügers Archiv*, **393**, 190–7.)

results of an experiment carried out by these authors during the course of a study on the electrical effects of adding transported solutes either to the peritubular capillaries or to the tubular lumen. In this instance 100 mM glucose was added to the peritubular fluid while the membrane potential across the brush border membrane was recorded. It is clear that when the intracellular concentration of solute rises (as a result of glucose entry via an electrically silent facilitated transport system in the basolateral membrane) the membrane at the apical surface of the epithelium hyperpolarises; this effect reverses upon removal of glucose. This shows clearly that the sodium-coupled glucose co-transporter present in the apical surface is sensitive not only to the electrochemical driving force for sodium ions across this membrane but also to the solute concentration on both sides of the apical membrane (see also Chapter 3). As a consequence of this general observation, the transporters for amino acids which are found in the basal membrane must be of importance in regulating not only the intra-epithelial concentration of the various amino acids but also the rate at which amino acids are reabsorbed from the glomerular filtrate across the apical membrane. As with the small intestine a full understanding of interactions between amino acids and the regulation of transport at this surface it not yet available, but studies using cultured cells (Rabito & Karish, 1982) together with experiments on vesicles and electrophysiological work (Samarzija & Fromter 1982) have contributed some understanding. Thus Na^+ dependence of gly uptake has been demonstrated at the basal surface in studies on isolated perfused tubules and Schafter & Williams (1985) suggest that the transporter involved may be of importance in subserving the maintenance of cellular amino acid pools necessary for metabolism, because the uptake of amino acids from the lumen is limited by the low glycine concentration found in the lumen of the late proximal tubule.

Basolateral membranes in general appear to transport amino acids by saturable systems which, contrary to the apical (brush border) membrane, are often Na^+ independent (see Table 5.2). A striking exception to this statement is found for glutamate (Sacktor *et al.*, 1981; see Figure 5.6) where an inwardly directed Na^+ gradient drives an overshoot in intravesicular glutamic acid concentration, and this overshoot is further enhanced by an outwardly directed K^+ gradient. Since this transport is electro-

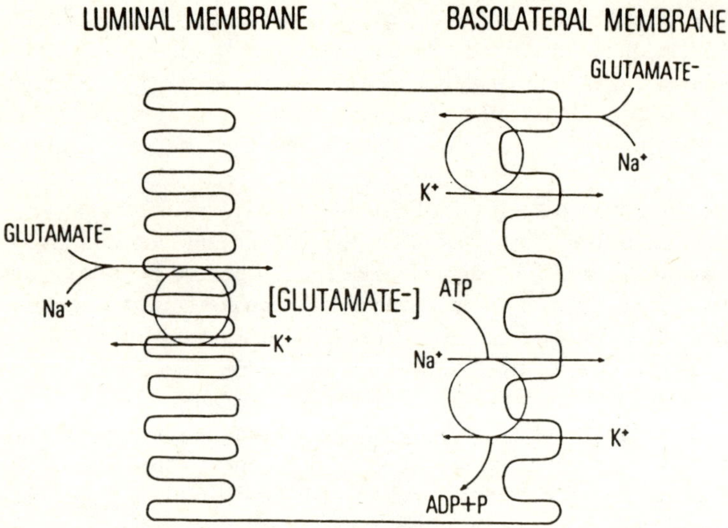

Fig. 5.6 Schematic model for L-glutamate uptake by the renal tubular cell. The concentrative uptake of L-glutamate energised by a Na^+ gradient and augmented by a K^+ gradient, at both sides of the polar tubular cell, accounts for the exceedingly high intracellular concentration of this metabolically important amino acid in the renal tubular epithelium. (Taken from Sacktor, B. *et al.* (1981). *Journal of Membrane Biology*, **60**, 63–71.)

genic *in vivo* the question of the stoichiometry of Na/K/glutamate transport remains unresolved, but cannot simply be 2:1:1. Since a similar system is found in the brush border, the lack of asymmetry of glutamate transport coupled to the multiple electrochemical gradients coupled to its movement must account for the very high intracellular concentrations of this amino acid which are built up in many tissues including kidney (cf. Table 5.2).

5.6 **Amino acid transport in the placenta**

The structural unit which forms the basis of placental transport of amino acids is the trophoblast which is an unusual type of epithelium since it separates two blood compartments, i.e. the maternal and fetal circulations (Figure 5.7). The trophoblast which forms a true syncytium (in which there are no lateral intercellular

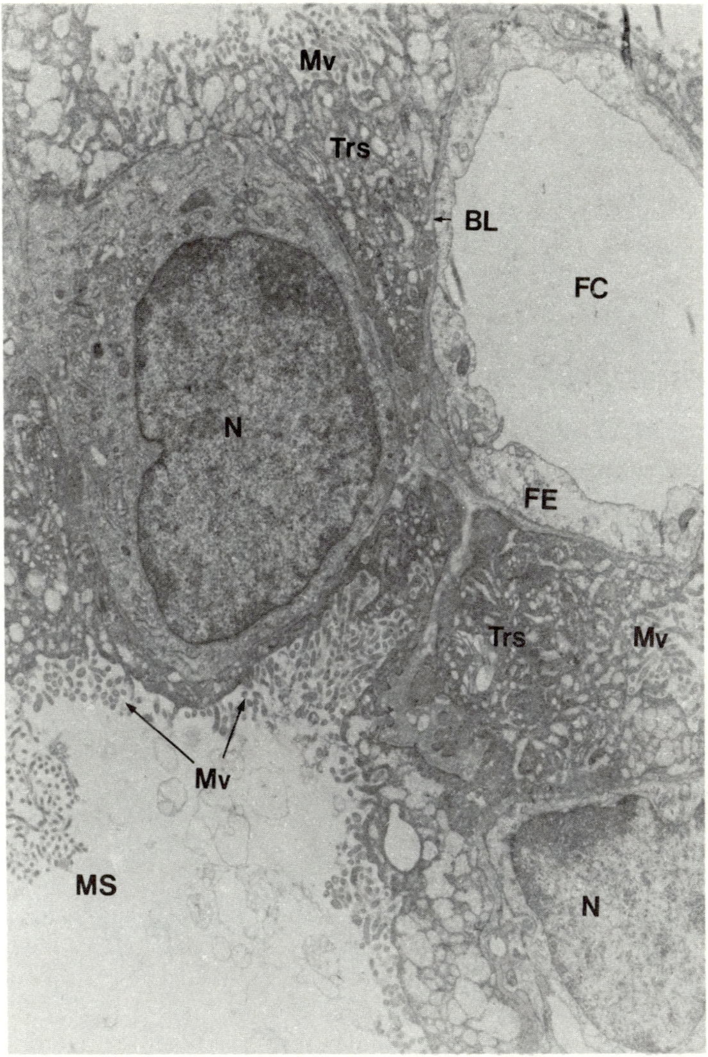

Fig. 5.7 Electron micrograph, of the guinea-pig hemomonochorial placenta. The fetal capillary (FC) and maternal blood (spaces (MS) are separated by the fetal capillary endothelium (FE) and the syncytiotrophoblast (Trs). It can be seen that the exchange barrier of the trophoblast may be very thin in certain areas. Mv, microvilli along the maternal border of the Trs. N, nucleus; BL, basal or fetal side of the Trs. (magnification × 8917). (Micrograph kindly provided by Dr G. Gabella, University College, London.)

spaces) has two poles: the maternal surface has microvilli which in structure resemble the brush border of the intestinal or renal epithelium whereas the fetal surface is equivalent to the basal side of such epithelia. Transport of amino acids from the maternal circulation will serve the high metabolic functions of the trophoblast; however most amino acids will be transferred rapidly across the fetal side to provide for fetal metabolic requirements (see review by Yudilevich & Sweiry, 1985).

Fetal/maternal ratios of amino acid plasma concentrations have been used as an index of uphill transport across the placenta ('transplacental' transport). It can be seen in Table 5.1 that these ratios vary from about 4 (for histidine) to close to unity (for example for aspartate and citrulline). As in most other tissues, individual amino acids are generally found in much higher concentrations in the trophoblast cytoplasm than in either of the extracellular compartments (Table 5.1). Hill & Young (1973 – see Yudilevich & Sweiry, 1985), in the fetal side perfused (closed circuit) guinea-pig placenta, demonstrated transfer from the maternal circulation to the fetal perfusate against a concentration gradient. Most amino acids are taken up from the maternal circulation (positive arterio-venous difference) whereas on the fetal side in general amino acids appear to be secreted into the fetal circulation thus producing negative arterio-venous differences. It is interesting to compare these differences at maternal and fetal sides of the placenta with those in the liver (positive uptake) and skeletal muscle (negative uptake), respectively (see Fig. 2 in Christensen, 1982).

In the singly and dually-perfused guinea pig placenta, Yudilevich and colleagues demonstrated and characterised membrane transporters which exist on both sides of the trophoblast (Yudilevich *et al.*, 1979; Yudilevich & Eaton, 1980; Eaton *et al.*, 1982). A typical experimental result with their paired-tracer technique is illustrated in Figure 5.8. Uptake of a range of amino acids (neutral, basic and acidic) was measured and for most amino acids high and similar uptakes were found at both surfaces of the trophoblast (Eaton & Yudilevich, 1981; see Table 2 in Yudilevich & Sweiry, 1985). The absence of a measurable uptake of taurine, glutamate or aspartate at either face of the placenta suggested that the trophoblast lacked both an anionic system and a system specific for β-amino acids. These authors also measured high and similar uptakes for lysine and arginine and a cationic transport system has now been characterised (Wheeler, 1986). In contrast to the human, in the guinea pig placenta System A is absent (or repressed) since (i) methyl-AIB (MeAIB) was not taken up whereas both alanine and AIB are transported, and (ii) there is no inhibitory effect of MeAla on the transport of alanine. The presence of an ASC-type carrier only was therefore proposed, supported by the fact that cysteine and other small neutral amino acids inhibit the uptake of alanine (Eaton *et al.*, 1982). The presence of System L was suggested by the inhibition of phenylalanine uptake by large neutral amino acids (methionine, leucine and phenylalanine) although some cross reactivity with the small neutral amino acids was also observed.

Kinetic constants for unidirectional influx of some natural amino acids were measured at the fetal and, separately, at the maternal interface (Yudilevich & Eaton, 1980; Eaton *et al.* 1982; Wheeler, 1986; see Figure 5.9 and Table 5.2). On the fetal side the K_m for alanine transport was similar to that on the maternal side. Interestingly sodium dependence is seen only at the maternal interface (Table 5.2). The K_m value for phenylalanine at the fetal side was three-fold higher than that on the maternal interface, suggesting different characteristics for the carriers at the two faces of the trophoblast. The presence of a high concentration of lysine (10 mM) did not effect phenylanine kinetics (Figure 5.9).

In the isolated dually-perfused placenta the efflux of amino acid from the tissue to either maternal or fetal circulations can be investigated. Figure 5.8 shows that uptake of lysine is followed by rapid backflux towards the ipsilateral circulation and that this backflux is significantly larger on the fetal side. In addition, there is transplacental transfer in both maternal-to-fetal and fetal-to-maternal directions although transplacental transfer (for all the amino acids which were taken up) was greater in the maternal-to-fetal direction. The asymmetry of the rate of efflux rather than the rate of entry at the two faces of the trophoblast may thus be seen as the mechanism underlying net transplacental transport of amino acids to the fetus (Eaton & Yudilevich, 1981; see Yudilevich & Sweiry, 1985).

Techniques for separating basal from brush border trophoblast

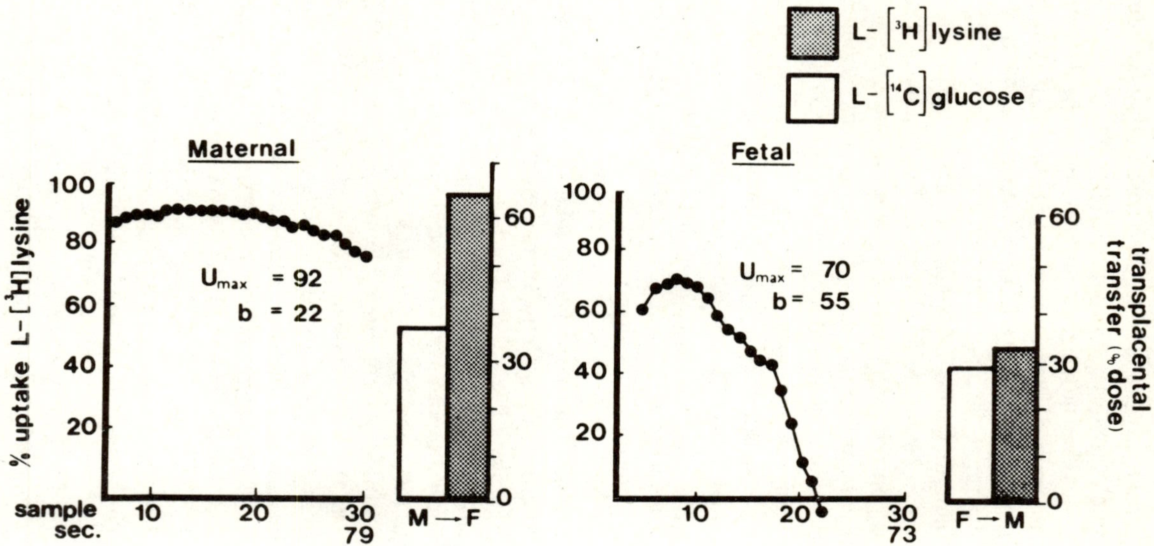

L- [3H] lysine
L- [14C] glucose
Maternal
Fetal
% uptake L-[3H]lysine
100
80
60
40
20
sample
sec.
10
20
30
79
73
Umax = 92
b = 22
Umax = 70
b = 55
M → F
F → M
60
30
0
transplacental
transfer (% dose)

membrane vesicles have been developed for human placenta but no reports have yet appeared on amino acid transport studies in vesicles isolated from the fetal side. Kinetic parameters in placental brush border membrane vesicles are given in Table 5.2. These studies showed that three systems exist for neutral amino acids: A, ASC and L (Enders *et al.*, 1976; see Smith in Kilberg, 1986; Yudilevich & Sweiry, 1985). The first (System A) group – glycine, proline and AIB – is strongly inhibited by MeAla but not by 2-aminonorborane-2-carboxylic acid (BCH). The second (System L) group – BCH, phenylalanine, valine and isoleucine – is inhibited strongly by BCH but very little by MeAla. The third (System ASC) group – alanine, serine, threonine and glutamine – is partially inhibited both by BCH and by MeAla. These results suggest that species difference exists between human and guinea-

---

Fig. 5.8 (opposite) Uptake of lysine at the brush-border (maternal) and basal (fetal) sides of the perfused placenta. Upper figure: L-[^{3}H]lysine uptakes versus time (s). A mixture of a labelled amino acid, L-[^{3}H]lysine, and a labelled extracellular marker, L-[^{14}C]glucose, was injected into either the maternal or fetal arterial inflow of the isolated dually-perfused guinea-pig placenta; sequential venous samples were collected immediately. Uptake in each sample was calculated as (1 − [^{3}H]lysine/[^{14}C] glucose) × 100. Maximal initial uptake values (U_{max}) were followed by a decrease in the 'apparent' uptake which was related to tracer backflux, *b*. The value of *b* (backflux over the 5–6 min collection period) is given as a percentage of U_{max}. The unidirectional influx into the trophoblast, *v*, could be estimated from: $v = -F \cdot \ln(1 - U_{max}) \cdot C$, where *F* is the rate of blood flow ($ml \cdot min^{-1} \cdot g^{-1}$) and *C* is the concentration of the unlabelled amino acid in the perfusate (Eaton *et al.*, 1982). From these unidirectional influx measurements, saturation kinetic parameters may be estimated (see Figure 5.9).

Transplacental transfer of the two-tracers was estimated from samples collected on the contralateral side; since L-[^{14}C]glucose transport is presumably via nonspecific pathways ('leakage'), the excess L-[^{3}H]lysine is a measure of the specific transfer of this amino acid. As may be seen net lysine transport is found since the maternal-to-fetal (M→F) flux is larger than the fetal-to-maternal (F→M) transfer. (Data taken from Eaton & Yudilevich (1981), and from unpublished results of D. L. Yudilevich, C. P. D. Wheeler, & J. H. Sweiry.)

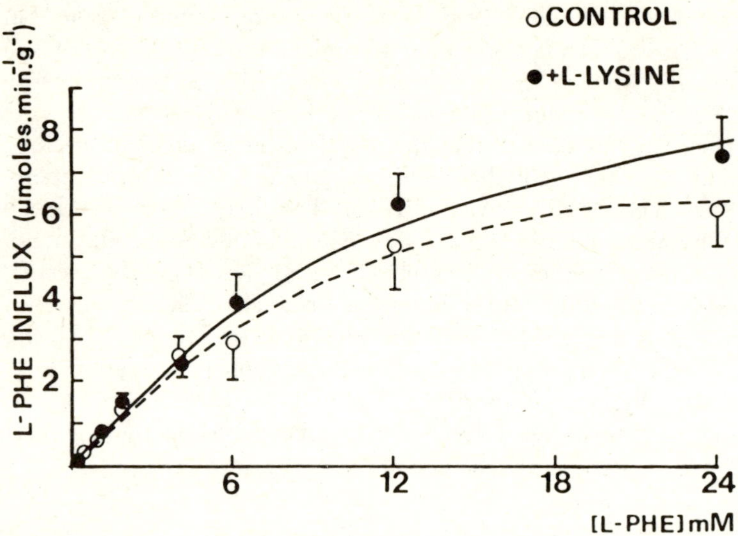

Fig. 5.9 Lack of effect of unlabelled L-lysine on transport kinetics of neutral amino acids at the placental fetal interface. L-phenylalanine (0.05–24.05 mM) saturation kinetics in the presence of 10 mM L-lysine, (n = 4 placentas). These experiments were performed in the singly-perfused guinea pig placenta and unidirectional influx was measured by means of the paired tracer technique (see Figure 5.8; Yudilevich & Eaton, 1980; Eaton *et al.*, 1982; Yudilevich, 1985). (Taken with permission from Wheeler, C. P. D. (1986). Ph.D. thesis, University of London.)

pig placentae since the latter appears to have only System ASC.

Boyd & Lund (1981) used human placental brush-border membrane vesicles to study the characteristics of proline uptake. They found that although amino acids usually associated with System L were inhibitory, glycine and MeAIB were the most effective competitors. The presence of System A was therefore suggested. The poorly hydrolysed dipeptide glycyl-L-proline inhibited proline uptake as effectively as proline itself suggesting that there may also be a transport system for oligopeptides which may compete directly or indirectly for the same electrochemical driving force. The apparent K_m for proline was similar to that reported for AIB (Table 5.2). Accumulation of L-proline within placental vesicles is electrogenic and linked to the movement of Na^+: when the relatively permeant chloride anion was replaced in the incubation medium by the relatively impermeant sulphate anion, the transient accumulation of L-proline within the vesicles was abolished (Figure 5.10). Moreover when the vesicles, again with chloride replaced by sulphate, were prepared with an outwardly directed K^+ gradient in the presence of valinomycin, L-proline was transiently accumulated within the vesicles (Figure 5.10). In addition, the accumulation of $^{24}Na^+$ appeared to be transiently stimulated by a gradient of L-proline but not D-proline, confirming the co-transport nature of the process.

Recently Ganapathy *et al.* (1986) have characterised the transport of tryptophan in human placental vesicles. They concluded that System T (see section 4.4.5) is not expressed in this tissue and that aromatic amino acids are transported via System L.

5.7 Amino acid transport in exocrine glands: salivary gland and pancreas

In exocrine organs uptake of amino acids from the blood occurs at the basolateral side of the epithelium but there is little transfer of free amino acid into the lumen. Accumulation inside the cell is required for the high rates of protein synthesis found in the cells of these organs.

Influx of a number of amino acids across the basolateral side of the salivary epithelium has been studied in the cat submandibular gland (Bustamante *et al.*, 1981; Mann & Yudilevich, 1984; Mann *et al.*, 1984). Figure 5.11 shows the results which can be summarized as follows:

(i) the presence of an ASC System is suggested by the uptake of alanine and serine, and by the lack of uptake of methyl-AIB and AIB – the latter also indicates the absence of an A System;
(ii) the minimal uptake of glycine and proline suggests the lack of the System Gly;
(iii) taurine uptake indicates the presence of β-System;
(iv) there is a system for basic amino acids;
(v) an acidic system is probably absent.

In cross-competition experiments and when sodium-free perfusates were used, further kinetic characterisation (Table 5.2)

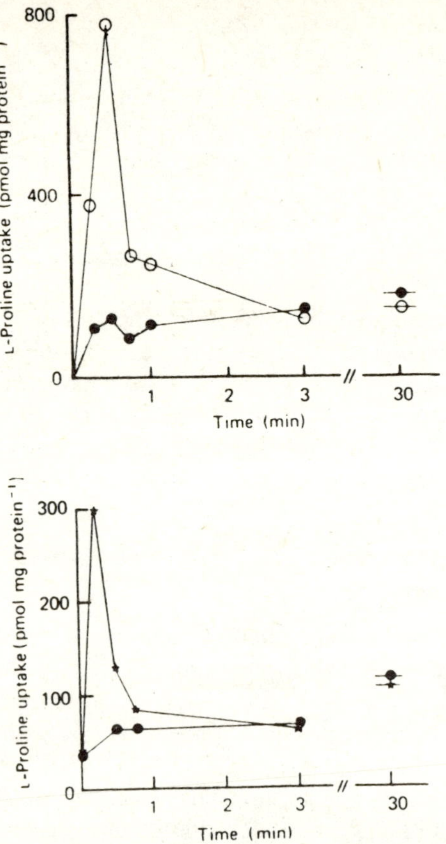

Fig. 5.10 Electrogenicity of placental brush-border membrane transport of proline. *Upper panel*: Effect on L-proline (0.25 mM) uptake by human placental brush-border vesicles of replacing chloride by sulphate in the incubation medium. When the relatively permeable chloride anion (100 mM) (○), was replaced by the relatively impermeable sulphate anion (100 mM), (•), the accumulation of proline within the vesicles was inhibited suggesting charge transfer by the amino acid/sodium cotransporter. *Lower panel*: The vesicles were prepared with a K^+ gradient (inside higher than outside), with equal Na^+ concentrations on both sides, and with chloride replaced by sulphate. Uptake of proline was measured in the presence (*) and absence (•) of valinomycin. In the presence of valinomycin (which increases K^+ conductance and therefore leads to an electrical potential difference, inside-negative, across the vesicle membrane) an increased uptake of proline is seen, suggesting that a net positive charge is transferred along with L-proline. (Taken with permission from Boyd, C. A. R. & Lund, E. K. (1981). *Journal of Physiology*, **315**, 9–19).

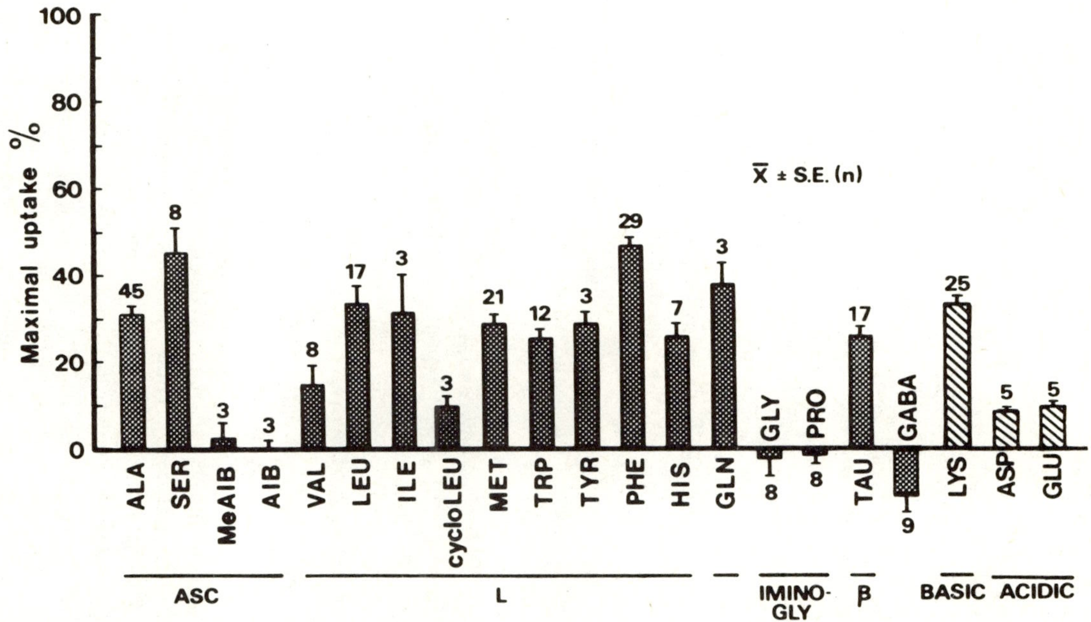

Fig. 5.11 Uptake of amino acids at the basolateral side of salivary epithelium. The paired-tracer method (Yudilevich & Mann, 1982; see Figure 5.8) was employed in the cat submandibular gland. For abbreviations see Table 5.2. Possible specific transport systems (see Table 1.1) are identified in the bottom of the figure. The maximal unidirectional uptake was measured for 18 natural amino acids and three non-metabolised synthetic analogues in glands with an intact blood supply (dotted columns) or in glands perfused with an amino acid-free Krebs-Henseleit solution (stippled columns). (Data taken from Mann, G. E. *et al*. (1984), and unpublished results of G. E. Mann, & D. L. Yudilevich, for the acidic amino acids.)

of the systems for neutral and basic amino acids was possible (Mann & Yudilevich, 1984; Mann *et al.*, 1984). Recently vesicles were obtained from the basolateral side of the rat parotid gland and uptake of AIB was used to kinetically (Table 5.2) characterize a System A which was inhibited by the use of the ionophore monensin (Takuma & Baum, 1985).

As distinct from the salivary epithelium, the existence of an A System was identified in the pancreas on the basis of the high rate of MeAIB uptake and from the fact that transport of this model amino acid was markedly affected by the sodium gradient (Table 5.2; see Mann & Peran, 1986). In both exocrine organs uptake of long-chain neutrals appeared to be via System L. However the pancreatic transport systems revealed a lower degree of selectivity than that observed in the salivary gland. Unlike the salivary epithelium, small neutral amino acids enter the pancreas via a Na^+ independent System asc (hormone and diabetes sensitive). Influx kinetics in both exocrine organs are shown in Table 5.2. Efflux of phenylalanine or serine from the pancreas is accelerated by increasing perfusate concentrations of certain amino acids. Electrophysiological studies have shown that neutral amino acids including L-alanine depolarise the pancreatic acinar cell basolateral membrane (Iwatsuki & Petersen, 1980a; see Chapter 3).

5.8 **Amino acid transport in the liver**

The important role of the liver in the metabolic handling of amino acids and in their interorgan distribution (Christensen, 1982) correlates with its avid uptake of these substrates from the blood (Kilberg, 1982; see Chapter 1 for detailed discussion of regulation of transport in the hepatocyte). Table 5.2 illustrates some recent kinetic data. The spectrum of kinetic parameters for sodium dependent AIB uptake (ranging from 1 to 50 $mmol \cdot l^{-1}$) has been discussed by Kilberg *et al.* (1985a) and the following comments concerning this variability are useful for a critical evaluation of kinetic estimates in general (i.e. for the appraisal of data in Table 5.2):

(i) Cell preparation used, i.e. freshly isolated hepatocytes or primary monolayer cultures;
(ii) cell viability;

(iii) length of time used for transport measurement;
(iv) the degree of gene expression;
(v) the contribution of various systems in the absence or presence of sodium.

A comparison between the kinetics of alanine and glutamate transport (Table 5.2), both performed in vesicles by the same group (see Sips *et al.*, 1982) indicates that the capacity (V_{max}) of the acidic system is two orders of magnitude lower than that for the neutral system. The affinity constants (K_m) for alanine and aspartate were 3.0 mM and 5.4 μM respectively. A recent finding is that histidine is transported at significant rates by System A as well as by System N (Bracy *et al.*, 1986). Notwithstanding these problems kinetic analysis has been useful: thus White (1985) showed that the glutamine-specific System N had the highest activity (V_{max}/K_m = 22 nmol · mg^{-1} protein · mM^{-1} followed by System-ASC (V_{max}/K_m = 2.8), System L (V_{max}/K_m = 1.4), System A (V_{max}/K_m = 0.66) and System y^+ (V_{max}/K_m = 0.28). Hepatocytes isolated from glucagon-treated rats show a five-fold stimulation of arginine influx measured after 2 h of culture (Handlogten & Kilberg, 1984; see White, 1985).

Interestingly the normal liver seems to lack a cationic system y^+ which is however active in hepatoma cells (see White, 1985). Among other notable 'absentees' is the lack of a system for anionic amino acids (see Table 1 in Christensen, 1982).

5.9 **Amino acid transport in the brain**

5.9.1 *Blood–brain barrier*

Transport of substrates at the blood–brain barrier (BBB) is important as a limiting factor of brain metabolism (see review by Pardridge, 1983). Even though some authors claim the existence in the brain of up to 10 amino acid transport systems others believe that the brain capillary is particularly lacking most of the systems represented elsewhere (Christensen, 1979). These contradictory opinions are related to the various cell types found in the brain and the methods employed to investigate them. The membrane transporters of the endothelium must be distinguished from those of other brain cells. Furthermore, we now understand that the luminal and antiluminal plasma membranes of the brain endothelial cell have distinct transport characteristics (see Betz & Goldstein, 1986). The luminal side has been studied in the brain *in situ* by means of methods which measure the rapid uptake from the blood, whereas studies on isolated brain capillaries can also give information on the antiluminal membrane (Betz & Goldstein, 1978; see Table 5.2 and Figure 5.12).

Yudilevich *et al.* (1972; 1975) measured rapid blood–tissue uptake of amino acids by dog brain *in vivo*. They found a large influx of large neutral amino acids (and histidine) whereas the influx for short-chain neutrals, acidics and basics was negligible. These authors proposed that the luminal side of the brain capillary may lack the widely distributed A System (as well as other specific carriers for this group) and that short-chain neutrals seem also to have very low affinity for the L System (see Table 5.2). Interestingly in their studies they found no transport of amino acids such as aspartate, glutamate and glycine, all three of which have been proposed to be neurotransmitters.

In the rat, Oldendorf (1971 – see Oldendorf, 1976), found rapid uptake for long-chain neutrals but low, although measurable, uptake for small neutrals and for basic amino acids (see Pardridge, 1983). Wade & Katzman (1975) using Oldendorf's technique further defined the existence of the L System measuring the uptake of BCH. Wade & Katzman (1975) confirmed the conclusion reached by Yudilevich *et al.* (1972) that the A System is absent in the BBB, since MeAIB had no measurable uptake and neither did glycine or other small neutrals that strongly prefer System A (see Christensen, 1979). Furthermore the lack of effect of insulin on the blood to brain transport of neutral amino acids (Daniel *et al.*, 1981) is consistent with the absence of System A.

Kinetic constants for amino acid transport at the BBB have been measured (Table 5.2). Recently a kinetic analysis of isoleucine transport in the *in situ* perfused rat brain has shown that the affinity is five-fold greater than previously measured with the Oldendorf technique (Takasato *et al.*, 1985). Many estimates of K_m determined for single amino acids in the BBB are however of little value in a physiological sense since the net affinity of the carrier system for a given amino acid varies depending on the

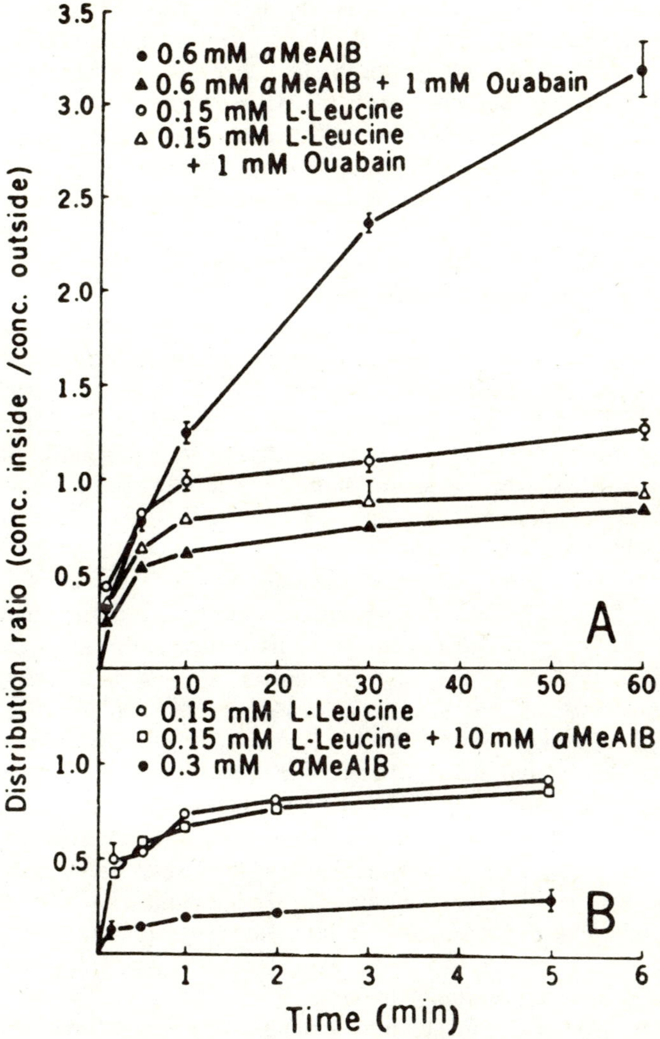

Fig. 5.12 Amino acid uptake of [^{14}C]MeAIB and l-[^{14}C]leucine in isolated brain capillaries. A. Isolated capillaries were first incubated for 20 min at 37°C with or without 1 mM ouabain. B. Isolated capillaries were prepared in NaCl buffer but then washed three times with a Na^+-free (choline substituted) medium. After 20 min of incubation at 37°C, uptake in Na^+-free medium was measured. (Taken with permission from Betz, A. C. & Goldstein, G. W. (1978). *Science*, **202**, 225–6. Copyright 1978 by the American Association for the advancement of Science).

plasma concentrations of competing amino acids. Transport competition has been discussed in relation to the importance for brain uptake (and metabolism) of the high levels of phenylalanine found in phenylketonuria (Pardridge, 1983; see Christensen 1986).

The proposal that there may be a System A type carrier located at the antiluminal side of the brain capillary therefore producing transport of the short-chain neutrals from the brain interstitium (hence also from the cerebrospinal fluid) towards the blood compartment, was supported by the finding (Figure 5.12) of transport of MeAIB in capillaries isolated from the rat brain (Betz & Goldstein, 1978). Interestingly, in capillaries isolated from rats subjected to experimental portacaval anastomosis it was found that the activity of System L, but not that of System A, was enhanced (Cardelli-Cangiano *et al.*, 1981; see Pardridge, 1983).

Christensen (1979) discusses the implications for the polarity of amino acid transport into and out of the brain if only one transport system (System L) plays a significant part in the movement of amino acids from the blood, and Systems L and A are present on the antiluminal side (Betz & Goldstein, 1978). Furthermore, Pardridge (1983) claims that glutamate transport at the antiluminal surface may be a way of disposing of this neurotransmitter from the brain interstitium into the circulation, a suggestion which would also require a transporter for efflux at the blood-side. Tryptophan movement from blood into brain has particular relevance due to its important role in the synthesis of serotonin. Even though the BBB seems to be able to make albumin-bound tryptophan available for transport into the brain (see Pardridge, 1983), the same does not happen in the liver since in isolated hepatocytes the addition of albumin to the medium considerably reduced the influx of tryptophan into the cell (Christensen, 1979).

5.9.2 *Blood–cerebrospinal fluid barrier*

The concentrations of most amino acids in the cerebrospinal fluid (CSF) are considerably lower than in the plasma. Transport of amino acids at the blood-CSF barrier (situated at the choroid plexus and other circumventricular organs) has been investigated by ventriculo-cisternal perfusion of labelled amino acids, in the rabbit (Davson *et al.*, 1982). These authors provided evidence of uphill transport from the CSF to blood and suggested that the normally low concentration of amino acids in the CSF is due to an active transport from CSF to blood (Davson *et al.*, 1982).

In the choroid plexus no net transport of amino acids across the choroidal epithelium from the blood into the cerebrospinal fluid was found (Wright, 1972 – see Wright, 1978). Wright proposed that the pia-arachnoid is 'a more reasonable candidate for active transport of amino acids from the CSF to blood'. Levin *et al.* (1974) showed that the pial capillary takes up amino acids very rapidly from the CSF and demonstrated a method which allows studies of transport across the antiluminal side of the pial capillary (a tight endothelium similar to the BBB).

5.9.3 *Brain tissue: axon, synaptosome, glia*

Transport systems for amino acids in such tissue may provide not only for their intracellular accumulation as nutrients but also may provide mechanisms of re-uptake and subsequent destruction of amino acids acting as synaptic neurotransmitters. Mediated uptake of glycine, glutamate and other amino acids was demonstrated in the non-dialysed axon by Baker & Potasher (1973), and Caldwell (1978), as reviewed by Baker & Carruthers (1984). The transporter for the neurotransmitter γ-aminobutyric acid (GABA) (which is found at a concentration of about 2 $mmol \cdot l^{-1}$ in rat brain tissue; Herbert *et al.*, 1966) exhibits an absolute dependance not only on Na^+ but specifically on Cl^-; the proposed stoichiometry is the influx of two sodium ions and one chloride ion per GABA-zwitterion (Kanner & Radian, 1985). In mouse cerebrum, measurements of AIB and GABA influx showed that the carriers for these amino acids appeared to be kinetically distinct (Cohen, 1985). The efflux of taurine and glycine (which do not competitively interact in most cells) was shown to be sodium dependent in internally dialysed giant axon (Horn, 1986). Glial cells have been isolated and transport of amino acids investigated in these preparations (Bowery *et al.*, 1979, Weiler *et al.*, 1979 – see Baker & Carruthers, 1984).

5.10 **Concluding remarks on polarity of transcellular amino acid transport**

Amino acid transport by the placenta resembles that seen in other epithelia with respect to the ability of the tissue to achieve net transepithelial transfer of amino acid. Differences in the density of pumps and leaks in the maternal (brush border) and fetal (basal) faces of the trophoblast provide an explanation for polarised transepithelial amino acid transport. For this tissue a realistic model would involve inwardly directed Na^+-coupled pumps at both faces with Na^+ independent carrier-mediated 'leaks' being found predominantly in the fetal-facing membrane. In the intestine and kidney, however, the asymmetry which produces net transport appears to arise more from the predominant location of Na^+-coupled transport in the brush border, although the possibility of an active exit mechanism for amino acids in the basolateral side has not been totally excluded (see discussion by Cheeseman, 1981; 1983). The blood–brain barrier formed by the capillary endothelium appears to be designed on similar principles with specific, powered amino acid transporters found on the antiluminal side.

Autoradiographic techniques have been employed to localise transported amino acids within intestinal (King *et al.*, 1981) and salivary (Mann *et al.*, 1986) epithelia. In the villous epithelium of the intestine accumulation of amino acids is greatest at the tips of the villi and this is believed to reflect a maturation process as migration time from crypt to villous tip takes about 2–4 days. In the salivary gland amino acids appeared to enter acinar, demilunar and striated ductal cells across the basolateral membrane without further transport of free amino acids into the salivary secretion, i.e. cellular uptake from the blood serves protein synthesis. The cellular resolution obtained using both light and electron microscopic autoradiography may in future permit alterations in amino acid transport to be correlated with a membrane defect specific to a single epithelial cell type.

Since many of the epithelial cells considered are highly active metabolically the inter-relationship in such tissues between amino acid transport and metabolism needs careful consideration (e.g. Christensen, 1983; McGivan *et al., 1981; Kilberg et al.*, 1985a). Table 5.4 illustrates for hepatocytes the relationship between the rates of gluconeogenesis and the rates of transport when differing amino acids are provided as substrates. This study highlights the essential interdependence of transport and metabolism. Both processes are of course catalysed by proteins and it is by the study

Table 5.4 *Comparison of the transport and gluconeogenic rates (pmol · mg^{-1} protein · min^{-1}) for selected amino acids in cultured rat hepatocytes*[a]

Amino acid tested	Transport velocity (T)	Gluconeogenic velocity (G)	Ratio, T/G[b]
Glycine	26	180	0.41
Cysteine	50	125	0.40
Alanine	50	100	0.50
Serine	75	140	0.54
Threonine	22	40	0.55
Proline	30	50	0.60
Asparagine	22	25	0.88
Methionine	60	25	2.40
Glutamine	250	50	5.00
Histidine	300	35	8.60
Phenylalanine	270	30	9.00

The rate of amino acid-dependent gluconeogenesis was estimated by measuring the incorporation of ^{14}C-labelled amino acid carbon into glucose (Claus *et al.*, 1975, in Kilberg *et al.* below). The initial rate of transport of the amino acid was measured in freshly isolated hepatocytes (McGiven *et al.*, 1977).

[a]Modified after Kilberg, M. S. *et al.* (1985). *Current Topics in Cellular Regulation*, 25, 133–161.

[b]T/G ratio provides an evaluation of which of the two processes is slower.

of the structure, mechanism, synthesis and control of such proteins that a detailed functional picture will be discovered. Oxender *et al.* (1985) have recently made a start in the molecular characterisation of the System L transporter. However for tissue function to be understood progress will also be needed in understanding how cells distribute amino acid transporters vectorially in their apical and basal plasma membranes and how the rates of synthesis and degradation of these transporters are matched to changes in metabolic and other physiological requirements.

5.11 **Further reading**

(Other literature is listed in the compiled References at the end of the book.)

Betz, A. L. & Goldstein, G. W. (1986). Specialized properties and solute transport in brain capillaries. *Annual Review of Physiology,* **48**, 241–50.

Boyd, C. A. R. & Perring, V. S. (1982). Amino acid inhibition and stimulation of α-aminoisobutyric acid exit from anuran small intestine. *Journal of Physiology*, **327**, 53–64.

Cheeseman, C. I. (1983). Characteristics of lysine transport across the serosal pole of the anuran small intestine. *Journal of Physiology*, **338**, 87–97.

Giebisch, G., *ed.* (1979). *Membrane transport in biology. Vol. IVB. Transport organs*, Springer-Verlag, Berlin.

Mann, G. E. & Yudilevich, D. L. (1984). Discrimination of parallel neutral amino acid uptake at the basolateral side of the epithelium in the cat salivary epithelium. *Journal of Physiology*, **347**, 111–27.

Murer, H. & Kinne, R. (1980). The use of isolated membrane vesicles to study epithelial transport processes. *Journal of Membrane Biology*, **55**, 81–95.

Murer, H. & Burckhardt, G. (1983). Transport of organic anions in epithelia. *Reviews in Physiology, Biochemistry and Pharmacology*, **96**, 2–38.

Schafer, J. A. & Williams, J. C. Jr. (1985). Transport of metabolic substrates by the proximal nephron. *Annual Review of Physiology*, **47**, 103–25.

Stevens, B. R. & Kaunitz, J. D. & Wright, E. M. (1984). Intestinal transport of amino acids and sugar; advances using membrane vesicles. *Annual Review of Physiology*, **46**, 417–33.

Yudilevich, D. L. & Sweiry, J. H. (1985). Transport of amino acids in the placenta. *Biochimica et Biophysica Acta*, **822**, 169–201.

References

(Selected Further Reading has been given at the end of each chapter)

Allemand, D., De Renzis, G., Ciapa, B., Girar, J.-P. & Payan, P. (1984). Characterization of valine transport in sea urchin eggs. *Biochimica et Biophysica Acta*, **772**, 337–46.

Al-Saleh, E. A. S. & Wheeler, K. P. (1982). Transport of neutral amino acids in human erythrocytes. *Biochimica et Biophysica Acta*, **684**, 157–71.

Alvarado, F. & Van Os, C. H. (1986). *25 years of research on the brush border membrane and sodium-coupled transport.* INSERM Symposium Series Vol. 26, Elsevier, Amsterdam.

Andreoli, T. E., Hoffman, J. F. & Fanestil, D. D. (1978). *Physiology of membrane disorders.* Plenunm, New York.

Appleman, J. R. & Lienhard, G. E. (1985). Rapid kinetics of the glucose transporter from human erythrocytes. Detection and measurement of a half-turnover of the purified transporter. *Journal of Biological Chemistry*, **260**, 4575–8.

Aronson, P. S. (1978). Energy-dependence of phlorizin binding to isolated renal microvillus membranes. Evidence concerning the mechanism of coupling between the electrochemical Na^+ gradient and sugar transport. *Journal of Membrane Biology*, **42**, 81–98.

Aull, F. (1967). Measurement of the electrical potential difference across the membrane of the Ehrlich mouse ascites tumor cell. *Journal of Cellular Physiology*, **69**, 21–32.

Baker, P. F. & Carruthers, A. (1984). Transport of sugars and amino acids. In: *The squid axon* (ed. P. F. Baker), Vol. 22, pp. 91–130. Current Topics in Membranes and Transport.

Bannai, S. (1984a). Transport of cystine and cysteine in mammalian cells. *Biochimica et Biophysica Acta*, **779**, 289–306.

Bannai, S. (1984b). Induction of cysteine and glutamate transport activity in human fibroblasts by diethyl maleate and other electrophilic agents. *Journal of Biological Chemistry*, **259**, 2435–40.

Bannai, S., Christensen, H. N., Vadgama, J. V., Ellory, J. C., Englesberg, E., Guidotti, G. G., Gazzola, G. C., Kilberg, M. S., Lajtha, A., Sacktor, B., Sepulveda, F. V., Young, J. D., Yudilevich, D. L. & Mann, G. E. (1984). Amino acids transport systems. *Nature*, **311**, 308.

Bannai, S. & Tateishi, N. (1986). Role of membrane transport in metabolism and function of glutathione in mammals. *Journal of Membrane Biology*, **89**, 1–8.

Baños, G., Daniel, P. M., Moorhouse, S. R., Pratt, O. E., Wilson, P. A., (1978). The influx of amino acids into the heart of the rat. *Journal of Physiology*, **280**, 471–86.

Barret, G. C. (1985). *Chemistry and biochemistry of the amino acids*, Chapman and Hall, London.

Benson, P. F. & Fensom, A. H. (1985). *Genetic biochemical disorders*. Oxford University Press, Oxford.

Betz, A. L. & Goldstein, G. W. (1978). Polarity of the blood-brain barrier: neutral amino acid transport into isolated brain capillaries. *Science*, **202**, 225–6.

Betz, A. L. & Goldstein, G. W. (1986). Specialized properties and solute transport in brain capillaries. *Annual Review of Physiology*, **48**, 241–50.

Boerner, P. & Saier, M. H., Jr. (1985). Adaptive regulatory control of System A transport activity in a kidney epithelial cell line (MDCK) and in a transformed variant (MDCK-T1). *Journal of Cellular Physiology*, **122**, 308–15.

Boge, G., Rigal, A. & Peres, G. (1983). Analysis of two chloride requirements for sodium-dependent amino acid and glucose transport by intestinal brush-border membrane vesicles of fish. *Biochimica et Biophysica Acta*, **729**, 209–18.

Bowery, N. G., Brown, D. A., White, R. D. & Yamini, G. (1979). [^{3}H] gamma-aminobutyric acid uptake into neuroglial cells of rat superior cervical sympathetic ganglia. *Journal of Physiology*, **293**, 51–74.

Boyd, C. A. R. (1979). Studies on amino acid inhibition of monosaccharide exit from anuran small intestinal epithelium. *Journal of Physiology*, **294**, 195–210.

Boyd, C. A. R., Cheeseman, C. I. & Parsons, D. S. (1975). Amino acid movements across the wall of anuran small intestine perfused through the vascular bed. *Journal of Physiology*, **250**, 409–29.

Boyd, C. A. R. & Lund, E. K. (1981). L-proline transport by brush border membrane vesicles prepared from human placenta. *Journal of Physiology*, **315**, 9–19.

Boyd, C. A. R. & Perring, V. S. (1981). Transamination and asymmetry in glutamate transport across the basolateral membrane of frog small intestine. *Bioscience Reports*, **1**, 851–56.

Boyd, C. A. R. & Perring, V. S. (1982). Amino acid inhibition and stimulation of 2-aminoisobutyric acid exit from anuran small intestine. *Journal of Physiology*, **327**, 53–64.

Boyd, C. A. R. & Ward, M. R. (1982). A micro-electrode study of oligopeptide absorption by the small intestinal epithelium of *Necturus maculosus*. *Journal of Physiology*, **324**, 411–28.

Bracy, D. S., Handlogten, M. E., Barber, E. F., Han, H.-P. & Kilberg, M. S. (1986). *Cis*-inhibition, *trans*-inhibition, and repression of hepatic amino acid transport mediated by system A. *Journal of Biological Chemistry*, **261**, 1514–20.

Britton, H. G. (1965). Fluxes in passive, monovalent and polyvalent carrier systems. *Journal of Theoretical Biology*, **10**, 28–52.

Brooker, R. J. & Wilson, T. H. (1985). Isolation and nucleotide sequencing of lactose carrier mutants that transport maltose. *Proceedings of the National Academy of Science*, U.S.A., **82**, 3959–63.

Brown, P. D. & Sepulveda, F. V. (1985). Potassium movements associated with amino acid and sugar transport in enterocytes isolated from rabbit jejunum. *Journal of Physiology*, **363**, 271–86.

Burckhardt, G., Kinne, R., Stange, G. & Murer, H. (1980). The effects of potassium and membrane potential on sodium-dependent glutamic acid uptake. *Biochimica et Biophysica Acta*, **599**, 191–201.

Bustamante, J. C., Mann, G. E. & Yudilevich, D. L. (1981). Specificity of neutral amino acid uptake at the basolateral side of the epithelium in the cat salivary gland *in situ. Journal of Physiology*, **313**, 65–79.

Cardelli-Cangiano, P., Cangiano, C., James, J. H., Jeppsson, B., Brenner, W. & Fischer, J. E. (1981). Uptake of amino acids by brain microvessels isolated from rats after portacaval anastomosis. *Journal of Neurochemistry*, **36**, 627–32.

Cheeseman, C. I. (1981). The mechanism of transfer for L-leucine into the vascular bed of the anuran small intestine. *Journal of Physiology*, **317**, 91–102.

Cheeseman, C. I. (1983). Characteristics of lysine transport across the serosal pole of the anuran small intestine. *Journal of Physiology*, **338**, 87–97.

Cheeseman, C. I., King, I. & Smith, M. W. (1983). Tissue distribution of exogenous amino acids during transport across the vascularly perfused anuran small intestine. *Journal of Physiology*, **339**, 133–44.

Chen, C.-C., Tsuchiya, T., Yamane, Y., Wood, J. M. & Wilson, T. H. (1985). Na^+ (Li^+)-proline cotransport in Escherichia coli. *Journal of Membrane Biology*, **84**, 157–64.

Chipperfield, A. R. & Perring, V. S. (1980). Furosemide-sensitive glycine transport in human erythrocytes. *Journal of Physiology*, **310**, 21P.

Christensen, H. N. (1953). Metabolism of amino acids and proteins. *Annual Review of Biochemistry*, **22**, 233–60.

Christensen, H. N. (1973). On the development of amino acid transport systems. *Federation Proceedings*, **32**, 19–28.

Christensen, H. N. (1975). *Biological Transport*, 2nd Edn. W. A. Benjamin, Reading, Mass.

Christensen, H. N. (1979). Developments in amino acid transport, illustrated for the blood-brain barrier. *Biochemical Pharmacology*, **28**, 1989–92.

Christensen, H. N. (1982). Interorgan amino acid nutrition, *Physiological Reviews*, **62**, 1193–1233.

Christensen, H. N. (1983). Hypothesis: Control of hepatic utilization of alanine by membrane transport or by cellular metabolism. *Bioscience Reports*, **3**, 905–13.

Christensen, H. N. (1984). Organic ion transport during seven decades. The amino acids. *Biochimica et Biophysica Acta*, **779**, 255–69.

Christensen, H. N. (1985). On the strategy of kinetic discrimination of amino acid transport systems. *Journal of Membrane Biology*, **84**, 97–103.

Christensen, H. N. (1986). Where the circulating amino acids go in PKU and why. *Biochemical Journal* (in press).

Christensen, H. N., de Cespedes, C., Handlogten, M. E. & Ronquist, G. (1973). Energization of amino acid transport, studied for the Ehrlich ascites tumour cell. *Biochimica et Biophysica Acta*, **300**, 487–522.

Christensen, H. N. & Clifford, J. B. (1963). Early postnatal intensification of hepatic acumulation of amino acids. *Journal of Biological Chemistry*, **238**, 1743–5.

Christensen, H. N., Streicher, J. A. & Elbinger, R. L. (1948a). Effects of feeding individual amino acids upon the distribution of other amino acids between cells and extracellular fluid. *Journal of Biological Chemistry*, **172**, 515–24.

Cohen, S. R. (1985). Rate equations and kinetics of uptake of α-aminoisobutyric acid and γ-aminobutyric acid by mouse cerebrum slices incubated in media containing L(+)-lactate or a mixture of succinate, L-malate, and pyruvate as the energy source. *Journal of Neurochemistry*, **44**, 455–64.

Corcelli, A., Prezioso, G., Palmieri, F. & Storelli, C. (1982). Electroneutral Na^+/dicarboxylic amino acid cotransport in rat intestinal brush border membrane vesicles. *Biochimica et Biophysica Acta*, **689**, 97–105.

Crane, R. K. (1962). Hypothesis for mechanism of intestinal active transport of sugars. *Federation Proceedings*, **21**, 891–5.

Crane, R. K. (1977). The gradient hypothesis and other models of carrier-mediated active transport. *Reviews in Physiology, Biochemistry and Pharmacology*, **78**, 99–159.

Curran, P. F., Schultz, S. G., Chez, R. A. & Fuisz, R. E. (1967). Kinetic relations of the Na-amino acid interaction at the mucosal border of intestine. *Journal of General Physiology*, **50**, 1261–86.

Czech, M. P. (1985). *Molecular basis of insulin action.* Plenum Press, London.

Dall'asta, V., Gazzola, G. C., Ranchi-Gazzola, R., Bussolati, O., Longo, N. & Guidotti, G. C. (1983). Pathways of L-glutamic acid transport in cultured human fibroblasts. *Journal of Biological Chemistry*, **259**, 6371–9.

Daniel, P. M., Love, E. R., Moorehouse, S. R. & Pratt, O. E. (1981). The effect of insulin upon the influx of tryptophan into the brain of the rabbit. *Journal of Physiology*, **312**, 511–62.

Danisi, G., Tai, Y-H. & Curran, P. F. (1976). Mucosal and serosal fluxes of alanine in rabbit ileum. *Biochimica et Biophysica Acta*, **455**, 200–13.

Davson, H., Hollingsworth, J. G., Carey, M. B. & Fenstermacher, J. D. (1982). Ventriculo-cisternal perfusion of twelve amino acids in the rabbit. *Journal of Neurobiology*, **13**, 293–318.

Dawson, W. D., Robinson, S. C. & Smith, T. C. (1983). The use of $HgCl_2$ to evaluate the cosubstrate: amino acid transport stoichiometry in Ehrlich ascites tumor cells. *Journal of Cellular Physiology*, **115**, 131–6.

Devés, R. & Krupka, R. M. (1984). The comparative specificity of the inner and outer substrate transfer sites in the choline carrier of the human erythrocytes. *Journal of Membrane Biology*, **80**, 71–80.

Eaton, B. M. & Yudilevich, D. L. (1981). Uptake and asymmetric efflux of amino acids at maternal and fetal sides of placenta. *American Journal of Physiology*, **241**, C106–C112.

Eaton, B. M., Mann, G. E. & Yudilevich, D. L. (1982). Transport specificity for neutral and basic amino acids at maternal and fetal interfaces of the guinea pig placenta. *Journal of Physiology*, **328**, 245–58.

Eddy, A. A. (1968a). A net gain of sodium ions and a net loss of potassium ions accompanying the uptake of glycine by mouse ascites-tumour cells in the pesence of sodium cyanide. *Biochemical Journal*, **108**, 195–206.

Eddy, A. A. (1968b). The effects of varying the cellular and extracellular concentrations of sodium and potassium ions on the uptake of glycine by mouse ascites-tumour cells in the presence and absence of sodium cyanide. *Biochemical Journal*, **108**, 489–98.

Eddy, A. A. (1969). A sodium ion concentration gradient formed during the absorption of glycine by mouse ascites-tumour cells. *Biochemical Journal*, **115**, 505–9.

Eddy, A. A. (1977). Current status of the ion gradient hypothesis in some selected systems. In: *Intestinal permeation* (eds. M. Kramer & F. Lauterbach), pp. 332–49. Excerpta Medica, Amsterdam.

Eddy, A. A. (1978). Proton-dependent solute transport in micro-organisms. *Current Topics in Membranes and Transport*, **10**, 279–359.

Eddy, A. A. (1985). Sodium cotransport systems and the membrane potential difference. *Annals of the New York Academy of Sciences*, **456**, 51–62.

Eddy, A. A., Mulcahy, M. F. & Thomson, P. J. (1967). The effects of sodium ions and potassium ions on glycine uptake by mouse ascites-tumour cells in the presence and absence of selected metabolic inhibitors. *Biochemical Journal*, **103**, 863–76.

Ellory, J. C., Jones, S. E. M., Preston, R. L. & Young, J. D. (1982). A high-affinity sodium-dependent transport system for glutamate in dog red cells. *Journal of Physiology*, **320**, 79P.

Ellory, J. C., Jones, S. E. M. & Young, J. D. (1981). Glycine transport in human erythrocytes. *Journal of Physiology*, **320**, 403–22.

Ellory, J. C. & Lew, V. L. (1977). *Membrane transport in red cells.* Academic Press, New York.

Ellory, J. C. & Osotemehin, B. (1983). Glutamine uptake in human erythrocytes. *Journal of Physiology*, **348**, 59P.

Ellory, J. C., Tucker, E. M. & Deversen, E. V. (1972). The identification of ornithine and lysine at high concentrations in the red cells of sheep with an inherited deficiency of reduced glutathione. *Biochimica et Biophysica Acta*, **279**, 481–3.

Ellory, J. C. & Young, J. D. (1982). *Red cell membranes. A methodological approach.* Academic Press, London.

Elwyn, D. H. (1966). Distribution of amino acids between plasma and red blood cells in the dog. *Federation Proceedings*, **25**, 854–61.

Epel, D. (1972). Activation of an Na^+-dependent amino acid transport system upon fertilization of sea urchin eggs. *Experimental Cellular Research*, **72**, 74–89.

Fafournoux, P., Remesy, C. & Demigne, C. (1983). Control of alanine metabolism in rat liver by transport processes or cellular metabolism. *Biochemical Journal*, **210**, 645–52.

Fafournoux, P., Remesy, C. & Demigne, C. (1985). Control by amino acids of the activity of system A-mediated amino acid transport in isolated rat hepatocytes. *Biochemical Journal*, **231**, 315–20.

Fincham, D. A., Mason, D. H. & Young, J. D. (1985). Characterisation of a novel Na-independent amino acid transporter in horse erythrocytes. *Biochemical Journal*, **227**, 13–20.

Fincham, D. A., Mason, D. K., Paterson, J. Y. F. & Young, J. D. (1987). Heterogeneity of amino acid transport in horse erythrocytes. A detailed kinetic analysis of inherited transport variation. *Journal of Physiology* (in press).

Fisher (1954). *Protein metabolism.* Methuen, London.

Fisher, R. B. (1964). The mutability of K_m. In: *Oxygen and the animal organism* (eds. E. Neil & R. Torrance) pp. 339–49. Blackwell, Oxford.

Freel, R. W. & Goldner, A. M. (1981). Sodium-coupled nonelectrolyte transport across epithelia: emerging concepts and directions. *American Journal of Physiology*, **241**, G451–G460.

Frizzell, R. A. & Schultz, S. G. (1970). Effects of monovalent cations on the sodium-alanine interaction in rabbit ileum. Implication of anionic groups in sodium binding. *Journal of General Physiology*, **56**, 462–90.

Fromter, E. (1979). Solute transport across epithelia: What can we learn from micropuncture studies on kidney tubules? *Journal of Physiology*, **288**, 1–31.

Fukuhara, Y. & Turner, R. J. (1984). The static head method of determining the charge stoichiometry of coupled transport systems. Application to the sodium-coupled D-glucose transporters of the renal proximal tubule. *Biochimica et Biophysica Acta*, **770**, 73–78.

Ganapathy, V., Burckhardt, G. & Leibach, H. (1985). Peptide transport in rabbit intestinal brush-border membrane vesicles studied with a potential-sensitive dye. *Biochimica et Biophysica Acta*, **816**, 234–40.

Ganapathy, M. E., Leibach, F. H., Mahesh, V. B., Howard, J. C., Devoe, L. D. & Ganapathy, V. (1986). Characterization of tryptophan transport in human placental brush-border membrane vesicles. *Biochemical Journal*, **238**, 201–8.

Gazzola, G. C., Dall'asta, V., Franchi-Gassola, R., Bussolati, O., Long, N. & Guidotti, G. G. (1984). Post-translational control by carrier availability of amino acid transport in fetal human fibroblasts. *Biochemical Biophysical Research Communications*, **120**, 172–8.

Gazzola, G. C., Dall'asta, V. & Guidotti, G. G. (1981). Adaptive regulation of amino acid transport in cultured human fibroblasts. Sites and mechanisms of action. *Journal of Biological Chemistry*, **256**, 3191–8.

Gibb, L. E. & Eddy, A. A. (1972). An electrogenic sodium pump as a possible factor leading to the concentration of amino acids by mouse ascites-tumour cells with reversed sodium ion concentration gradients. *Biochemical Journal*, **129**, 979–81.

Giordana, B., Parenti, P., Hanozet, G. M. & Sacchi, V. F. (1985). Electrogenic K^+-basic amino acid cotransport in the midgut of Lepidopteran larvae. *Journal of Membrane Biology*, **84**, 46–53.

Greitz, T., Ingvar, D. H. & Widen, L. (1985). *Metabolism of the human brain studied with position emission tomography*. Raven Press, New York.

Grinstein, S. & Erlij, D. (1977). Differences between resting and insulin-stimulated amino acid transport in frog skeletal muscle. *Journal of Membrane Biology*, **35**, 9–28.

Guidotti, G. G., Borghetti, A. F. & Gazzola, G. G. (1978). The regulation of amino acid transport in animals cells. *Biochimica et Biophysica Acta*, **515**, 329–66.

Gunter-Smith, P. J., Grasset, E. & Schultz, S. G. (1982). Sodium-coupled amino acid and sugar transport by *Necturus* small intestine: an equivalent circuit analysis of a rheogenic co-transport system. *Journal of Membrane Biology*, **66**, 25–39.

Hacking, C. & Eddy, A. A. (1981). The accumulation of amino acids by mouse ascites-tumour cells. Dependence on but lack of equilibrium with the sodium-ion electrochemical gradient. *Biochemical Journal*, **194**, 415–26.

Handlogten, M. E., Barber, E. F., Bracy, D. S. & Kilberg, M. S. (1986). Amino acid-dependent inactivation of glucagon-induced System A transport activity in cultured rat hepatocytes. *Molecular and Cellular Endocrinology*, **43**, 61–69.

Handlogten, M. E. & Kilberg, M. S. (1984). Induction and decay of amino acid transport in the liver. Turnover of transport activity in isolated hepatocytes after stimulation by diabetes or glucagon. *Journal of Biological Chemistry*, **259**, 3519–25.

Hanson, P. J. & Parsons, D. S. (1986). Transport of glutamate in the small intestine. In: *Glutamine and glutamate metabolism* (ed. E. Kramme), pp. 1–35. CRC Pres Inc., Cleveland, Ohio.

Hardison, W. G. M. & Weiner, R. (1980). Taurine transport by rat hepatocytes in primary culture. *Biochimica et Biophysica Acta*, **598**, 145–52.

Haussinger, D., Soboll, S., Meijer, A. J., Gerok, W., Tager, J. M. & Sies, H. (1985). Role of plasma membrane transport in hepatic glutamine metabolism. *European Journal of Biochemistry*, **152**, 597–603.

Hayes, M. R. & McGivan, J. D. (1983). Comparison of the effects of certain thiol reagents on alanine transport in plasma membrane vesicles from rat liver and their use in identifying the alanine carrier. *Biochemical Journal*, **214**, 489–95.

Heinz, A., Jackson, J. W., Richey, B. E., Sachs, G. & Schafer, J. A. (1981). Amino acid active transport and stimulation by substrates in the absence of Na^+ electrochemical potential gradient. *Journal of Membrane Biology*, **62**, 149–60.

Heinz, E. (1978). *Mechanisms and energetics of biological transport.* Springer-Verlag, Berlin.

Heinz, E., Geck, P. & Wilbrandt, W. (1972). Coupling in secondary active transport. Activation of transport by co-transport and/or counter-transport with the fluxes of other solutes. *Biochimica et Biophysica Acta*, **255**, 442–61.

Heinz, E. & Patlak, C. S. (1960). Energy expenditure by active transport mechanisms. *Biochimica et Biophysica Acta*, **44**, 324–34.

Herbert, J. D., Coulson, R. A., & Hernandez, T. (1966). Free amino acids in the caiman and rat. *Comparative Biochemistry and Physiology*, **17**, 583–98.

Hill, T. L. (1982). The linear Onsager coefficients for biochemical kinetic diagrams as equilibrium one-way cycle fluxes. *Nature*, **299**, 84–86.

Hoare, D. (1972). The transport of L-leucine in human erythrocytes. A new kinetic analysis. *Journal of Physiology*, **221**, 311–29.

Hoffman, E. K. & Lambert, I. H. (1983). Amino acid transport and cell volume regulation in Ehrlich ascites tumour cells. *Journal of Physiology*, **338**, 613–25.

Holden, J. T. (1962). *Amino acid pools.* Elsevier, Amsterdam.

Hopfer, U. & Liedtke, C. M. (1981). Kinetic features of cotransport mechanisms under isotope exchange conditions. *Membrane Biochemistry*, **4**, 11–29.

Hopfer, U., Sigrist-Nelson, K., Ammann, E. & Murer, H. (1976). Differences in neutral amino acid and glucose transport between brush border and basolateral plasma membrane of intestinal epithelial cells. *Journal of Cell Physiology*, **89**, 805–10.

Horn, L. W. (1986). Measurements of amino acid transport in internally dialyzed giant axons. *Journal of Membrane Biology*, **89**, 185–92.

Hoshi, T. (1976). Electrophysiological studies on amino acid transport across the luminal membrane of the proximal tubular cells of *Triturus* kidney. In: *Amino acid transport and uric acid transport* (eds. S. Silbernagl, F. Lang & R. Greger), pp. 96–104. Georg Thieme, Stuttgart.

Hoshi, T. (1985). Proton-coupled transport of organic solutes in animal cell membranes and its relation to Na^+ transport. *Japanese Journal of Physiology*, **35**, 179–91.

Hudson, R. L. & Schultz, S. G. (1984). Sodium-coupled sugar transport: Effects on intracellular sodium activities and sodium-pump activity. *Science*, **224**, 1237–9.

Inui, K-I., Okano, T., Takano, M., Saito, H. & Hori, R. (1984). Carrier-mediated transport of cephalexin via the dipeptide transport system in rat renal brush-border membrane vesicles. *Biochimica et Biophysica Acta*, **769**, 449–54.

Inui, K., Saito, H. & Hori, R. (1985). H^+ gradient-dependent active transport of tetraethylammonium cation in apical-membrane vesicles isolated from kidney epithelial cell line LLC-PK_1. *Biochemical Journal*, **227**, 199–203.

Iwatsuki, N. & Petersen, O. H. (1980a). Amino acids evoke short-latency membrane conductance increase in pancreatic acinar cells. *Nature*, **283**, 492–94.

Iwatsuki, N. & Petersen, O. H. (1980b). Amino acid-evoked membrane potential and resistance changes in pancreatic acinar cells. *Pflügers Archiv*, **386**, 153–9.

Iwatsuki, N. & Petersen, O. H. (1985). Action of tetraethylammonium on calcium activated potassium channels in pig pancreatic acinar cells studied by patch-clamp single channel and whole-cell current recording. *Journal of Membrane Biology*, **86**, 139–44.

James, J. H., Jeppson, B., Ziparo, V. & Fischer, J. E. (1979). Hyperammonemia, plasma amino acid imbalance and blood-brain amino acid transport: a unified theory of portal systemic encephalopathy. *Lancet*, **2**, 772–5.

Johnstone, R. M. (1979). Electrogenic amino acid transport. *Canadian Journal of Physiology and Pharmacology*, **57**, 1–15.

Jung, D., Schwarz, W. & Passow, H. (1984). Sodium-alanine cotransport in oocytes of *Xenopus laevis*: correlation of alanine and sodium fluxes with potential and current changes. *Journal of Membrane Biology*, **72**, 29–34.

Kanner, B. I. & Radian, R. (1985). Ion-coupled neurotransmitter transport across the synaptic plasma membrane. In: *Membrane transport driven by ion gradients*, (eds. G. Semenza & R. Kinne). *Annals New York Academy of Science*, **456**, 153–161.

Kaunitz, J. D. & Wright, E. M. (1984). Kinetics of sodium D-glucose cotransport in bovine intestinal brush border vesicles. *Journal of Membrane Biology*, **79**, 41–51.

Kilberg, M. S. (1982). Amino acid transport in isolated rat hepatocytes. *Journal of Membrane Biology*, **69**, 1–33.

Kilberg, M. S. (ed.) (1986). Amino acid transport in eukaryotic cells and tissues. *Federation Proceedings*, **45**, 2438–54.

Kilberg, M. S., Barber, E. F. & Handlogten, M. E. (1985a). Characteristics and hormonal regulation of amino acid transport system A in isolated rat hepatocytes. *Current Topics in Cellular Regulation*, **25**, 133–61.

Kilberg, M. S., Han, H.-P., Barker, E. F. & Chiles, T. C. (1985b). Adaptive regulation of neutral amino acid transport system A in rat hepatoma cells. *Journal of Cellular Physiology*, **122**, 290–98.

Kilberg, M. S., Handlogten, M. E. & Christensen, H. N. (1980). Characteristics of an amino acid transport system in rat liver for glutamine, asparagine, histidine and closely related analogs. *Journal of Biological Chemistry*, **255**, 4011–19.

Kimmich, G. A. (1981). Intestinal absorption of sugar. In: *Physiology of the gastro-intestinal tract* (ed. L. R. Johnson), pp. 1035–61. Raven Press, New York.

Kimmich, G. A. & Randles, J. (1984). Sodium-sugar coupling stoichiometry in chick intestinal cells. *American Journal of Physiology*, **247**, C74–C82.

King, G. F. & Kuchel, P. (1984). A proton NMR study of iminopeptide transport and hydrolysis in the human red cell: possible physiological roles for the coupled system. *Biochemical Journal*, **220**, 553–60.

King, G. F. & Kuchel, P. (1985). Assimilation of α-glutamyl peptides by human erythrocytes: a possible means of glutamate supply for glutathione biosynthesis. *Biochemical Journal*, **227**, 833–42.

King, I. S., Sepulveda, F. V. & Smith, M. W. (1981). Cellular distribution of neutral and basic amino acid transport systems in rabbit ileal mucosal epithelium. *Journal of Physiology*, **319**, 355–68.

Kipnis, D. M. & Noall, M. W. (1958). Stimulation of amino acid transport by insulin in the isolated rat diaphragm. *Biochimica et Biophysica Acta*, **28**, 226–7.

Kletzien, R. F., Pariza, M. W., Becker, J. E., Potter, V. R. & Butcher, F. R. (1976). Induction of amino acid transport in primary cultures of adult rat liver parenchymal cells by insulin. *Journal of Biological Chemistry*, **251**, 3014–20.

Klip, A., Logan, W. J. & Gagalong, E. (1982). Regulation of amino acid transport in L6 myoblasts. II. Different chemical properties of transport after amino acid deprivation. *Journal of Cellular Physiology*, **113**, 56–66.

Kopito, R. R. & Lodish, H. F. (1985). Primary structure and transmembrane orientation of the murine anion exchange protein. Nature, **316**, 234–8.

Kostyo, J. L. (1968). Rapid effects of growth hormone on amino acid transport and protein synthesis. *Annals of the New York Academy of Sciences*, **148**, 389–407.

Kristensen, L. O. & Folke, M. (1983). Coupling ratio of electrogenic Na^+– alanine cotransport in isolated rat hepatocytes. *Biochemical Journal*, **210**, 621–4.

Kristensen, L. O., Sestoft, L. & Foulke, M. (1983). Concentrative uptake of alanine in hepatocytes in fed and fasted rats. *American Journal of Physiology*, **244**, G491–G500.

Lash, L. H. & Jones, D. P. (1984). Characteristics of cysteine uptake in intestinal basolateral membrane vesicles. *American Journal of Physiology*, **247**, G394–G401.

Lauger, P. (1980). Kinetic properties of ion carriers and channels. *Journal of Membrane Biology*, **57**, 163–78.

Laugier, R. & Petersen, O. H. (1981). Two different types of electrogenic amino acid action on pancreatic acinar cells. *Biochimica et Biophysica Acta*, **641**, 216–21.

LeCam, A. & Freychet, P. (1976). Glucagon stimulates the A system for neutral amino acid transport in isolated hepatocytes of adult rat. *Biochemical and Biophysical Research Communications*, **72**, 893–901.

LeCam, A. & Freychet, P. (1978). Effect of insulin on amino acid transport in isolated rat hepatocytes. *Diabetologia*, **15**, 117–23.

Lerner, J. (1977). *A review of amino acid transport processes in animal cells and tissues*. University of Maine at Orono Press, Maine.

Lerner, J. (1985). Effectors of amino acid transport processes in animal cell membranes. *Comparative Biochemistry and Physiology*, **81A**, 713–39.

Lever, J. E. (1980). The use of membrane vesicles in transport studies. *CRC Critical Reviews in Biochemistry*, **7**, 187–246.

Levin, E., Sepulveda, F. V. & Yudilevich, D. L. (1974). Pial vessels transport of substances from cerebrospinal fluid to blood. *Nature*, **249**, 266–8.

Lobitz, C. J. & Neville, M. C. (1977). Control of amino acid transport in the mammary gland of the pregnant mouse. *Journal of Supramolecular Structure*, **6**, 355–62.

McGiven, J. D., Ramsell, J. C. & Lacey, J. H. (1981). Stimulation of alanine transport and metabolism by cAMP in the hepatocyte from fed rats. Transport as rate limiting step for metabolism. *Biochimica et Biophysica Acta*, **644**, 295–304.

Maede, V., Inaba, M. & Taniguchi, N. (1982). Increase of Na-K ATPase activity, glutamate and aspartate uptake in dog erythrocytes associated with hereditary high accumulation of GSH, glutamate, glutamine, and aspartate. *Blood*, **61**, 493–9.

Makowske, M. & Christensen, H. N. (1982). Contrasts in transport systems for anionic amino acids in hepatocytes and a hepatoma cell line HTC. *Journal of Biological Chemistry*, **257**, 5663–70.

Mann, G. E., Moller, M., Hedemark Poulsen, J., Wilson, S. M., & Yudilevich, D. L. (1986). Autoradiographic localization of transported neutral amino acids in epithelia of cat submandibular and sublingual salivary glands. *Cell and Tissue Research*, **246**, 353–8.

Mann, G. E. & Peran, S. (1986). Basolateral amino acid transport systems in the perfused exocrine pancreas: sodium-dependency and kinetic interactions between influx and efflux mechanisms. *Biochimica et Biophysica Acta*, **858**, 263–74.

Mann, G. E. & Yudilevich, D. L. (1984). Discrimination of parallel neutral amino acid uptake at the basolateral side of the epithelium in the cat salivary epithelium. *Journal of Physiology*, **347**, 111–27.

Mann, G. E., Wilson, S. M. & Yudilevich, D. L. (1984). Characteristics of a cationic amino acid transport system in the basolateral membrane of the cat salivary epithelium. *Journal of Physiology*, **351**, 123–34.

Marty, A. & Neher, E. (1983). Tight-seal whole-cell recording. In: *Single-channel recording* (eds. B. Sakmann, & E. Neher), pp. 107–22. Plenum Press. New York.

Maruyama, Y., Petersen, O. H., Flanagan, P. & Pearson, G. T. (1983). Quantification of Ca^{2+}-activated K^+ channels under hormonal control in pig pancreas acinar cells. *Nature*, **302**, 228–32.

Matthews, D. M. & Payne, J. W. (1980). Transmembrane transport of small peptides. *Current Topics in Membranes and Transport*, **14**, 331–425.

McKusick, V. A. (ed.) (1983). *Catalogue of Mendelian inheritance in man*, 6th Edn., John Hopkins Press, Baltimore.

Mircheff, A. K., van Os, C. H. & Wright, E. M. (1980). Pathways for alanine transport in intestinal basal lateral membrane vesicles. *Journal of Membrane Biology*, **52**, 83–92.

Misfeldt, D. S. & Sanders, M. J. (1982). Transepithelial transport in cell culture: Stoichiometry of Na-phlorizin binding and Na/D-glucose cotransport. A two-step, two-sodium model of binding and translocation. *Journal of Membrane Biology*, **70**, 191–8.

Mitchell, P. (1963). Molecule, group and electron translocation through natural membranes. In: *The structure and function of the membranes and surfaces of cells* (eds. D. J. Bell, & J. K. Grant), pp. 142–169. Cambridge University Press, Cambridge.

Mitchell, P. (1977). Epilogue: from energetic abstraction to biochemical mechanism. In: *Microbial energetics* (eds. B. A. Haddock & W. A. Hamilton), pp. 383–423. Cambridge University Press, Cambridge.

Moffett, J. & Englesberg, E. (1984). Recessive constitutive mutant Chinese hamster ovary cells (CHO-K1) with an altered A system for amino acid transport and the mechanism of gene regulation of the A system. *Molecular and Cellular Biology*, **4**, 799–808.

Moller, J. V. & Sheikh, M. I. (1983). Renal organic anion tranport system: Pharmacological, physiological, and biochemical aspects. *Pharmacological Reviews*, **34**, 315–58.

Moore, P. A., Hayme, D. W. & Oxender, D. L. (1977). A role for aminoacyl-tRNA synthetases in the regulation of amino acid transport in mammalian cell lines. *Journal of Biological Chemistry*, **252**, 7427–30.

Moran, A., Turner, R. J. & Handler, J. S. (1983). Regulation of sodium-coupled glucose transport by glucose in a cultured epithelium. *The Journal of Biological Chemistry*, **258**, 15087–90.

Mueckler, M., Caruso, C., Baldwin, S. A., Panico, M., Blench, I., Morris, H. R., Allard, W. J., Lienhard, G. E. & Lodish, H. F. (1985). Sequence structure of a human glucose transporter. *Science*, **229**, 941–5.

Munck, B. G. (1981). Intestinal absorption of amino acids. In: *Physiology of the Gastro-intestinal tract* (ed. L. R. Johnson), pp. 1097–1122. Raven Press, New York.

Munro, H. N. (1970). *Mammalian protein metabolism*, vol. 1–4. Academic Press, London.

Murer, H. & Hopfer, U. (1976). Studies with membrane vesicles of polarity of intestinal epithelial amino acid transport. *Journal of Cellular Physiology*, **89**, 805–10.

Murer, H. & Kinne, R. (1977). Sidedness and coupling of transport processes in small intestinal and renal epithelia. In *Biochemistry of membrane transport* (eds. G. Semenza & E. Carafoli). FEBS Symposium 42, pp. 292–304, Springer-Verlag, Berlin.

Murer, H. & Kinne, R. (1980). The use of isolated membrane vesicles to study epithelial transport processes. *Journal of Membrane Biology*, **55**, 81–95.

Murer, H. & Burckhardt, G. (1983). Transport of organic anions in epithelia. *Reviews in Physiology, Biochemistry and Pharmacology*, **96**, 2–38.

Nadziejko, C. E. & Reichberg, S. B. (1984). Inhibition of sodium-independent amino acid transport by dexamethasone in rat hepatoma cells. *Journal of Biological Chemistry*, **259**, 394–8.

Naftalin, R. J. (1984). The thermostatics and thermodynamics of cotransport. *Biochimica et Biophysica Acta*, **778**, 155–75.

Nelson, P. J. & Rudnick, G. (1982). The role of chloride ion in platelet serotonin transport. *Journal of Biological Chemistry*, **257**, 6151–5.

Neville, M. C., Lobitz, C. J., Ripoll, E. A. & Tinney, C. (1980). The sites for δ-aminoisobutyric acid uptake in normal mammary gland and ascites tumour cell. A comparative study of mouse tissue *in vitro*. *Journal of Biological Chemistry*, **255**, 7311–16.

Noall, M. W., Riggs, T. R., Walker,. L. M. & Christensen, H. N. (1957). Endocrine control of amino acid transfer. Distribution of an unmetabolizable amino acid. *Science*, **126**, 1002–5.

Ohsawa, M., Kilberg, M. S., Kimmel, G. & Christensen, H. N. (1980). Energization of amino acid transport in energy-depleted Ehrlich cells and plasma membrane vesicles. *Biochimica et Biophysica Acta*, **599**, 175–90.

Oldendorf, W. H. & Szabo, J. (1976). Amino acid assignment to one of the three blood-brain barrier amino acid carriers. *American Journal of Physiology*, **230**, 94–98.

Oxender, D. L. (1965). Stereospecificity of amino acid transport for Ehrlich tumour cells. *Journal of Biological Chemistry*, **240**, 2976–82.

Oxender, D. L. & Christensen, H. N. (1963). Distinct mediating systems for the transport of neutral amino acids by the Ehrlich cell. *Journal of Biological Chemistry*, **238**, 3686–99.

Oxender, D. L., Collarini, E. J., Shotwell, M. A., Lobeton, C. D., Moreno, A. & Campbell, G. S. (1985). Regulation and genetics of amino acid transport. In: *Membrane transport driven by ion gradients* (eds. G. Semenza & R. Kinne). *Annals of the New York Academy of Sciences*, **456**, 387–97.

Pardridge, W. M. (1977). Kinetics of competitive inhibition of neutral amino acid transport across the blood-brain barrier. *Journal of Neurochemistry*, **28**, 103–8.

Pardridge, W. M. (1983). Brain metabolism: a perspective from the blood-brain barrier. *Physiological Reviews*, **63**, 1481–535.

Pariza, M. W., Kletzien, R. F., Butcher, F. R. & Potter, V. R. (1976). Inductions by hormones added singly, simultaneously or sequentially: what cultured hepatocytes can tell about metabolic regulation in the whole animal. *Advances in Enzyme Regulation*, **14**, 103–15.

Paterson, J. Y. F., Sepulveda, F. V. & Smith, M. W. (1980). Stoichiometry versus coupling ratio in the cotransport of Na and different neutral amino acids. *Biochimica et Biophysica Acta*, **603**, 288–97.

Peck, E. J. Jr. (1980). Receptors for amino acids. *Annual Review of Physiology*, **42**, 615–27.

Peerce, B. E. & Wright, E. M. (1985). Evidence for tyrosyl residues at the Na^+ site on the intestinal Na^+/glucose cotransporter. *Journal of Biological Chemistry*, **260**, 6026–31.

Petersen, O. H. (1976). Electrophysiology of mammalian gland cells. *Physiological Reviews*, **56**, 535–77.

Petersen, O. H. (1980). *Electrophysiology of gland cells.* Academic Press, London.

Petersen, O. H. (1985). Importance of electrical cell-cell communication in secretory epithelia. In: *Gap junctions* (eds. M. V. L. Bennett & D. Spray), pp. 315–24. Cold Spring Harbor Laboratory, New York.

Petersen, O. H. & Maruyama, Y. (1984). Calcium-activated potassium channels and their role in secretion. *Nature*, **307**, 693–6.

Petersen, O. H. (1986). Electrophysiology of acinar cells. In: *The exocrine pancreas: Biology, pathobiology and diseases* (ed. V. L. W. Go), pp. 141–61. Raven Press, New York.

Philo, R. D. & Eddy, A. A. (1978a). The membrane potential of mouse ascites-tumour cells studied with the fluorescent probe 3,3′-dipropyloxadicarbocyanine. *Biochemical Journal*, **174**, 801–10.

Philo, R. D. & Eddy, A. A. (1978b). Equilibrium and steady-state models of the coupling between the amino acid gradient and the sodium electrochemical gradient in mouse ascites-tumour cells. *Biochemical Journal*, **174**, 811–17.

Quay, S. C., Kline, E. L. & Oxender, D. L. (1975). Role of leucyl tRNA synthetase in regulation of branched-chain amino acid transport. *Proceedings of the National Academy of Science*, **72**, 3921–4.

Rabito, C. A. & Karish, M. V. (1982). Polarized amino acid transport by an epithelial cell line of renal origin (LLC-PK1). The basolateral systems. *Journal of Biological Chemistry*, **257**, 6802–8.

Racker, E. (1985). *Reconstitution of transporters, receptors and patho-*

logical states. Academic Press, London.

Racker, E., Resnick, R. J. & Feldman, R. (1985). Glycolysis and methyl aminoisobutyrate uptake in rat-1 cells transfected with ras or myc oncogenes. *Proceedings of the National Academy of Sciences U.S.A.*, **82**, 3535–8.

Rajantie, J., Simell, O. & Perheentupa, J. (1980). Basolateral-membrane transport defect for lysine in lysinuric protein intolerance. *Lancet*, **1**, 1219–21.

Reid, M., Gibb, L. E. & Eddy, A. A. (1974). Ionophore-mediated coupling between ion fluxes and amino acid absorption in mouse ascites-tumour cells. Restoration of the physiological gradients of methionine by valinomycin in the absence of adenosine triphosphate. *Biochemical Journal*, **140**, 383–93.

Riggs, T. R. (1970). Hormones and transport across cell membranes. In: *Biochemical actions of hormones* (ed. G. Litwack), pp. 157–208, Academic Press, New York.

Riggs, T. R., Christensen, H. N. & Palatine, I. M. (1952). Concentrating activity of reticulocytes for glycine. *Journal of Biological Chemistry*, **194**, 53–55.

Riggs, T. R., Pote, K. G., Im, H. S. & Huff, D. W. (1984). Thyroxine-induced changes in the development of neutral alpha-amino acid transport systems of rat brain. *Journal of Neurochemistry*, **42**, 1260–68.

Riggs, T. R., Walker, L. M. & Christensen, H. N. (1958). Potassium migration and amino acid transport. *Journal of Biological Chemistry*, **233**, 1479–84.

Rose, R. C. & Schultz, S. G. (1971). Studies on the electrical potential profile across rabbit ileum. Effects of sugars and amino acids on transmural and transmucosal electrical potential differences. *Journal of General Physiology*, **57**, 639–63.

Rosenberg, R. (1981a). L-leucine transport in human red blood cells. A detailed kinetic analysis. *Journal of Membrane Biology*, **62**, 79–93.

Rosenberg, R. (1981b). A kinetic analysis of L-tryptophan transport in human red blood cells. *Biochimica et Biophysica Acta*, **649**, 262–8.

Rosenberg, R., Young, J. D. & Ellory, J. C. (1980). L-tryptophan transport in human red blood cells. *Biochimica et Biophysica Acta*, **598**, 375–84.

Ruzycki, S. M., Kelley, L. K. & Smith, C. H. (1978). Placental amino acid uptake. IV. Transport by microvillous membrane vesicles. *American Journal of Physiology*, **234**, C27–C35.

Sacktor, B., Rosenbloom, I. L., Liang, C. T. & Cheng, L. (1981). Sodium gradient- and sodium plus potassium gradient-dependent L-glutamate uptake in renal basolateral membrane vesicles. *Journal of Membrane Biology*, **60**, 63–71.

Salter, M., Knowles, R. G. & Pogson, C. J. (1986). Transport of the aromatic amino acids into isolated liver cells. *Biochemical Journal*, **233**, 499–506.

Samarzija, I. & Fromter, E. (1982). Electrophysiological analysis of rat renal sugar and amino acid transport. III Neutral amino acids. *Pflügers Archiv*, **393**, 199–209.

Schafer, J. A. & Heinz, E. (1971). The effects of reversal of Na^+ and K^+ electrochemical potential gradients on the active transport of amino acids in Ehrlich ascites tumor cells. *Biochimica et Biophysica Acta*, **249**, 15–33.

Schafer, J. A. & Jacquez, J. A. (1967). Change in Na^+ uptake during amino acid transport. *Biochimica et Biophysica Acta*, **135**, 1081–83.

Schafer, J. A. & Williams, Jr. J. C. (1985). Transport of metabolic substrates by the proximal nephron. *Annual Review of Physiology*, **47**, 103–25.

Schell, R. E., Stevens, B. R. & Wright, E. M. (1983). Kinetics of sodium-dependent solute transport by rabbit renal and jejunal brush-border vesicles using a fluorescent dye. *Journal of Physiology*, **335**, 307–18.

Schneider, E. G. & Sacktor, B. (1980). Sodium gradient-dependent L-glutamate transport in renal brush border membrane vesicles. Effect of an intravesicular extravesicular potassium gradient. *Journal of Biological Chemistry*, **255**, 7645–9.

Schultz, I. & Ullrich, K. A. (1979). Transport processes in the exocrine pancreas. In: *Membrane transport in biology, Vol. IVB. Transport organs* (ed. G. Giebisch), pp. 811–852. Springer-Verlag, Berlin.

Schultz, S. G. (1977). Sodium-coupled solute transport by small intestine: a status report. *American Journal of Physiology*, **233**, E249–E254.

Schultz, S. G. & Curran, P. F. (1970). Coupled transport of sodium and organic solutes. *Physiological Reviews*, **50**, 637–718.

Schultz, S. G., Curran, P. F., Chez, R. A. & Fuisz, R. E. (1967). Alanine and sodium fluxes across mucosal border of rabbit ileum. *Journal of General Physiology*, **50**, 1241–1260.

Schultz, S. G. & Zalusky, R. (1964). Ion transport in isolated rabbit ileum. II The interaction between active sodium and active sugar transport. *Journal of General Physiology*, **47**, 1043–59.

Scriver, C. R. & Tenenhouse, H. S. (1985). Genetics and mammalian transport systems. In: *Membrane transport driven by ion gradients*, (eds. G. Semenza & R. Kinne). *Annals of the New York Academy of Sciences*, **456**, 384–97.

Semenza, G., Kessler, M., Hosang, M., Weber, J. & Schmidt, U. (1984). Biochemistry of the Na^+, D-glucose cotransporter of the small intestinal brush-border membrane. The state of the art in 1984. *Biochimica et Biophysica Acta*, **779**, 343–79.

Sepulveda, F. V. & Pearson, J. D. (1985). Cationic amino acid transport by two renal epithelial cell lines: LLC-PK1 and MDCK cells. *Journal of Cell Physiology*, **123**, 144–50.

Shotwell, M. A., Kilberg, M. S. & Oxender, D. L. (1983). The regulation of neutral amino acid transport in mammalian cells. *Biochimica et Biophysica Acta*, **737**, 267–284.

Sigrist-Nelson, K., Murer, H. & Hopfer, U. (1975). Active alanine transport in isolated brush border membranes., *Journal of Biological Chemistry*, **250**, 5674–80.

Singh, J. (1984). L-alanine and L-phenylalanine activate Na^+ and K^+ conductance pathways in the exocrine mouse pancreas. *Pflügers Archiv*, **402**, 176–84.

Singh, J. & Petersen, O.H. (1984). The effects of L-alanine and acetylcholine on membrane potential, $^{45}Ca^{2+}$ and $^{86}Rb^+$ efflux and amylase secretion in the isolated mouse pancres. *Quarterly Journal of Experimental Physiology*, **69**, 531–40.

Sips, H.J., DeGroof, P.A. & Van Dam, K. (1982). Transport of L-aspartate and L-glutamate in plasma membranes from the liver. *European Journal of Biochemistry*, **122**, 259–64.

Smith, T.C. (1982). The use of fluorescent dyes to measure membrane potentials: a response. *Journal of Cellular Physiology*, **112**, 302–5.

Stein, W.D. (1986). *Transport and diffusion across cell membranes.* Academic Press, New York.

Stevens, B.R., Kaunitz, J.D. & Wright, E.M. (1984). Intestinal transport of amino acids and sugar; advances using membrane vesicles. *Annual Review of Physiology*, **46**, 417–33.

Stevens, B.R. & Wright, E.M. (1985). Substrate specificity of the intestinal brush-border proline-sodium (IMINO) transporter. *Journal of Membrane Biology*, **87**, 27–34.

Takada, A. & Bannai, S. (1984). Transport of cystine in isolated rat hepatocytes in primary culture. *Journal of Biological Chemistry*, **259**, 2441–5.

Takasato, Y., Momma, S. & Smith, Q.R. (1985). Kinetic analysis of cerebrovascular isoleucine transport from saline and plasma. *Journal of Neurochemistry*, **45**, 1013–20.

Takuma, T. & Baum, B.J. (1985). Na^+-dependent transport of α-aminoisobutyrate in isolated basolateral membrane vesicles from rat parotid glands. *Biochimica et Biophysica Acta*, **812**, 453–9.

Takuwa, N., Shimada, T., Matsumoto, H. & Hoshi, T. (1985). Proton-coupled transport of glycylglycine in rabbit renal brush-border membrane vesicles. *Biochimica et Biophysica Acta*, **814**, 186–90.

Twitty, V.C. & Van Wagtendonk, W.J. (1940). A suggested mechanism for the regulation of proportionate growth supported by quantitative data on the blood nutrients. *Growth*, **4**, 349–50.

Tyrakowski, T.F., Milutinovic, S. & Schulz, I. (1978). Studies on isolated subcellular components of cat pancreas. III. Alanine-sodium co-transport in isolated plasma membrane vesicles. *Journal of Membrane Biology*, **38**, 333–46.

Ullrich, K.J., Rumrich, G. & Kloss, S. (1974). Sodium dependence of amino acid transport in the proximal convolution of the rat kidney. *Pflügers Archiv fur Gesamte Physiologie*, **351**, 49–60.

Ullrich, K.J. (1979). Sugar, amino acid and sodium cotransport in proximal tubule. *Annual Review of Physiology*, **41**, 181–96.

Vadgama, J. V. & Christensen, H. N. (1983). Comparison of System N in fetal hepatocytes and in related cell lines. *Journal of Biological Chemistry*, **258**, 6422–9.

Vadgama, J. V. & Christensen, H. N. (1985). Discrimination of Na^+-independent transport systems L, T, and asc in erythrocytes: Na^+-independence of the latter a consequence of cell maturation. Journal of Biological Chemistry, **260**, 2912–21.

Valdeolmillos, M., Garcia-Sancho, J. & Herreros, B. (1982). Stimulation of Na^+ dependent amino acid uptake by activation of the Ca^{2+}-dependent K^+ channel in the Ehrlich ascites tumor cell. *Biochimica et Biophysica Acta*, **689**, 177–9.

Van Slyke, D. D. & Meyer, G. M. (1913–1914). The fate of protein digestion products in the body. III. The absorption of amino-acids from the blood by the tissues. *Journal of Biological Chemistry*, **16**, 197–212.

Van Winkle, L. J., Christensen, H. N. & Campione, A. L. (1985). Na^+-dependent transport of basic, zwitterionic and bicyclic amino acids by a broad-scope system in mouse blastocysts. *Journal of Biological Chemistry*, **260**, 12118–23.

Vidaver, G. A. (1964). Some tests of the hypothesis that the sodium-ion gradient furnishes the energy for glycine-active transport by pigeon red cells. *Biochemistry*, **3**, 803–8.

Vidaver, G. A. & Shepherd, S. L. (1968). Transport of glycine by hemolyzed and restored pigeon red blood cells. *Journal of Biological Chemistry*, **243**, 6140–50.

Wade, L. A. & Katzman, R. (1975). Synthetic amino acids and the nature of L-DOPA transport at the blood-brain barrier. *Journal of Neurochemistry*, **25**, 837–42.

Wallach, D. F. H. (1979). *Plasma membranes and disease*. Academic Press, London.

Weigensberg, M. & Blostein, R. (1985). Na^+-coupled glycine transport in reticulocyte vesicles of distinct sidedness: Stoichiometry and symmetry. *Journal of Membrane Biology*, **86**, 37–44.

Weissbach, L., Handlogten, M. E., Christensen, H. N. & Kilberg, M. S. (1982). Evidence for two Na^+-independent neutral amino acid transport systems in primary cultures of hepatocytes. *Journal of Biological Chemistry*, **257**, 12006–11.

Weissbach, L. & Kilberg, M. S. (1984). Amino acid-activation of amino acid transport system N early in primary cultures of rat hepatocytes. *Journal of Cellular Physiology*, **121**, 133–8.

Wellner, D. & Meister, A. (1981) A survey of inborn errors of amino acid metabolism and transport in man. *Annual Review of Biochemistry*, **50**, 911–68.

Wheeler, K. P. & Christensen, H. N. (1967). Interdependent fluxes of amino acids and sodium ion in the pigeon red blood cell. *Journal of Biological Chemistry*, **242**, 3782–8.

Wheeler, C. P. D. (1986). *Transport of glucose, amino acids and adenosine in the perfused guinea-pig placenta*. Ph.D. Thesis, University of London.

White, J. T. & Armstrong, W. McD. (1971). Effect of transported solutes on membrane potentials in bullfrog small intestine. *American Journal of Physiology*, **221**, 194–201.

White, M. F. (1985). The transport of cationic amino acids across the plasma membrane of mammalian cells. *Biochimica et Biophysica Acta*, **822**, 355–74.

White, M. F. & Christensen, H. N. (1982). Cationic amino acid transport into cultured animals cells. *Journal of Biological Chemistry*, **257**, 4450–7.

White, M. F. & Christensen, H. N. (1983). Simultaneous regulation of amino acid influx and efflux by system A in the hepatoma cell HTC. Ouabain stimulates the starvation-induced derepression of System A amino acid transport. *Journal of Biological Chemistry*, **258**, 8028–38.

Widdas, W. F. (1980). The asymmetry of the hexose transfer system in the human red cell membrane. *Current Topics in Membrane Transport*, **14**, 165–223.

Wilson, T. H. & Wiseman, G. (1954). The use of sacs of everted small intestine for the study of the transference of substances from the mucosal to the serosal surface. *Journal of Physiology*, **123**, 116–25.

Wong, J. T.-F. (1975). *Kinetics of enzyme mechanisms*. Academic Press, London.

Wool, I. G., Castles, J. J. & Moyer, A. N. (1965). Regulation of amino acid accumulation in isolated rat diaphragm: effect of puromycin and insulin. *Biochimica et Biophysica Acta*, **107**, 333–45.

Wright, E. M. (1978). Transport processes in the formation of the cerebrospinal fluid. *Review of Physiology, Biochemistry, and Pharmacology*, **83**, 1–34.

Wright, J. K., Seckler, R. & Overath, P. (1986). Molecular aspects of sugar: ion cotransport. *Annual Review of Biochemistry*, **55**, 225–48.

Yamamoto, S.-I. & Kawasaki, T. (1981). The involvement of the membrane oxidoreduction system in stimulating amino acid uptake in Ehrlich ascites tumor cells. *Biochimica et Biophysica Acta*, **644**, 192–200.

Young, J. D. & Ellory, J. C. (1977). Red cell amino acid transport. In: *Membrane transport in red cells* (eds. J. C. Ellory & V. L. Lew), pp. 301–325. Academic Press, New York.

Young, J. D., Ellory, J. C. & Tucker, E. M. (1975). Amino acid transport defect in glutathione-deficient sheep erythrocytes. *Nature*, **254**, 156–7

.Young, J. D., Jones, S. E. M. & Ellory, J. C. (1980). Amino acid transport in human and sheep erythrocytes. *Proceedings of the Royal Society of London B*, **209**, 355–75.

Young, J. D., Jones, S. E. M. & Ellory, J. C. (1981). Amino acid transport via the red cell anion transport system. *Biochimica et Biophysica Acta*, **645**, 157–60.

Yudilevich, D. L. (1985). Characterization of membrane carriers and receptors of the blood-tissue interface by single circulation paired-tracer dilution. In: *Carrier mediated transport of solutes from blood to tissue* (eds. D. L. Yudilevich & G. E. Mann). Longman, London.

Yudilevich, D. L. & Mann, G. E. (eds.) (1985). *Carrier mediated transport of solutes from blood to tissue*. Longman, London.

Yudilevich, D. L., De Rose, N. & Sepulveda, F. V. (1972). Facilitated transport of amino acids through the blood brain barrier of the dog studied in a single capillary circulation. *Brain Research*, **44**, 569–78.

Yudilevich, D. L., Eaton, B. M., Short A. H. & Leichtweiss, H. P. (1979). Glucose carriers at the maternal and fetal sides of the trophoblast in guinea-pig placenta. *American Journal of Physiology*, **237**, C205–C212.

Yudilevich, D. L. & Eaton, B. M. (1980). Amino acid carriers at maternal and fetal surfaces of the placenta by single circulation paired tracer dilution. Kinetics of phenylalanine transport. *Biochimica et Biophysica Acta*, **595**, 315–19.

Yudilevich, D. L. & Mann, G. E. (1982). Unidirectional uptake of substrates at the blood-side of secretory epithelia: stomach, salivary gland, pancreas. *Federation Proceedings*, **41**, 3045–53.

Yudilevich, D. L. & Sepulveda, F. V. (1975). The specificity of amino acid and sugar carriers in the capillaries of the dog brain studied in vivo by rapid indicator dilution. In: *Transport phenomena in the nervous system* (eds. G. Levi, G. Battistin, & A. Lajtha), pp. 77–87. Plenum Press, New York.

Yudilevich, D. L. & Sweiry, J. H. (1985). Transport of amino acids in the placenta. *Biochimica et Biophysica Acta*, **822**, 169–201.

Index